CONTEMPORARY BRAINTEASERS

CONTEMPORARY BRAINTEASERS

TERRY STICKELS

DOVER PUBLICATIONS, INC.
MINEOLA, NEW YORK

Copyright

Bibliographical Note

Contemporary Brainteasers is a new work, first published by Dover Publications Inc., in 2016.

International Standard Book Number
ISBN-13: 978-0-486-80782-9
ISBN-10: 0-486-80782-7

Manufactured in the United States by LSC Communications
80782701 2016
www.doverpublications.com

CONTENTS

INTRODUCTION

I have read numerous articles on how puzzles are good exercise for the mind, how they augment critical thinking skills, and even how they can help in preventing or delaying dementia in some people. To date, I have not seen any scientific studies backing these positions. Of course, it would be to my advantage if this were otherwise. But there is a definite positive side to this. Without question, the better math puzzles offer the solver some interesting perspectives on how to solve problems. Puzzles are not bound by the same rigorous rules and algorithms of academic mathematics. Because of this, there is more wiggle room on how to reach a solution. In other words, solve the puzzle any way you can — we don't need to see your work to give you credit. But the mere fact that you are seeing solutions that you would never see in a classroom is a bonus in how you might approach other problems. Without any hard evidence, I make the claim that these types of puzzles have the ability to increase your mental flexibility. If you do enough of them, you will start to view other problems from different perspectives. Unfortunately, my position is anecdotal. There is no study confirming this. But if the views of my readers over the years are any indication, there is something to this and it stands to reason. If you show me five different ways to fix something or solve a problem, I am going to consider alternative approaches in other situations if I get stumped again.

The nice thing about puzzles is that they are for fun. No one is grading you or writing a job performance review. Kick back and try your skill at solving a few of these puzzles. It's in your blood. You are the direct result of many successful solutions made prior to your being here. Push those limits. We don't think anything about asking our young athletes to push their limits physically. Doesn't it make good sense that pushing the limits of your gray matter might have significant benefits?

The puzzles you now hold in your hands will push the limits of even the best puzzle solvers. But take that as a challenge to confront. Make these puzzles fun and see how good it feels when you solve one. (They are even more fun to create.) In any event, don't take yourself too seriously while solving these. Remember, challenging fun is all I ever intended.

PUZZLES

1

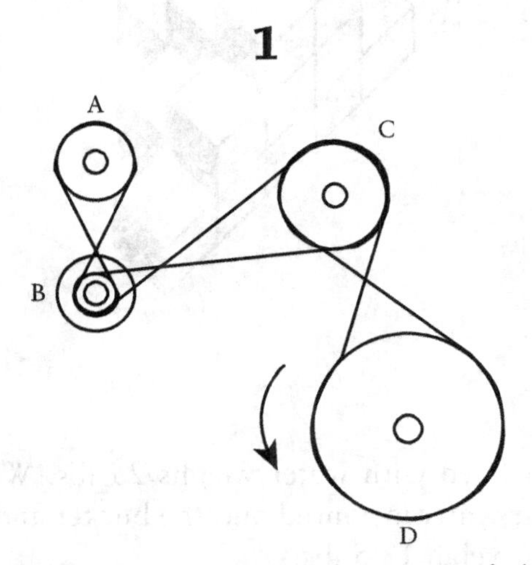

If pulley D is rotating in the direction of the arrow, then pulley A
(a) is moving in the same direction
(b) is moving in the opposite direction
(c) will not be able to move at all.

2

How many cubes of any size are contained in this stack? All rows and columns run to completion unless you actually see them end.

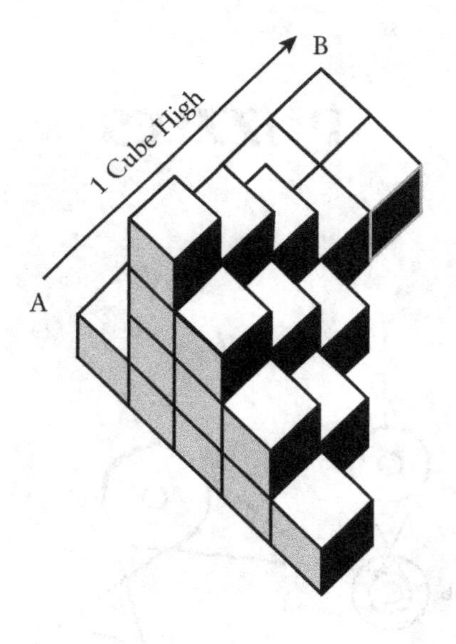

3

A bucket filled with water weighs 25 lbs. When one-half of the water is poured out, the bucket and remaining water weigh 13.5 lbs.

How much does the bucket weigh?

4

The three sides of a triangle have prime number lengths, w, x, and y, where $10<w<x<y<40$. If $x = 19$, how many possible perimeters does this triangle have?
(a) 5
(b) 8
(c) 13
(d) 21
(e) infinite

5

With a normal deck of 52 playing cards, on what card dealt is the probability of getting a straight (five consecutive cards of any suit) 100%?

In other words, how many cards would I have to deal to you, face down, before you tell me to stop because you know, with 100% accuracy, you have a straight?

6

If it were three hours later it would be half as long until midnight as it would be if it were one hour earlier. What time is it now?

7

If 1/3 = 7 in some other system, then 5/16 is equal to what in that system?

8

Each figure contained within the largest square is also a square. What are the sizes of the squares with the question marks?

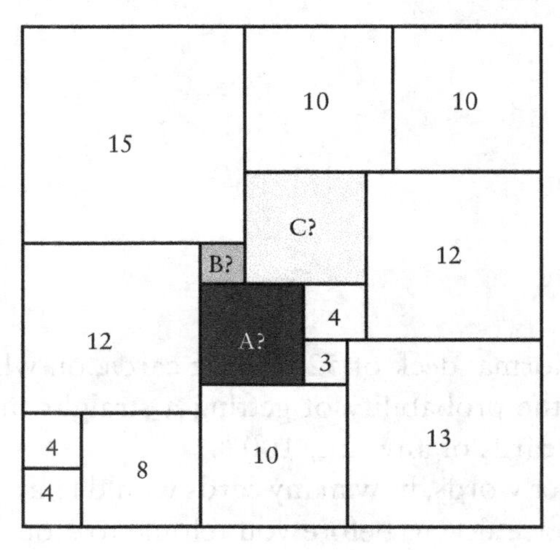

9

If the figure below is folded back into a cube, what side is opposite the blackened square? What about opposite the transparent circle?

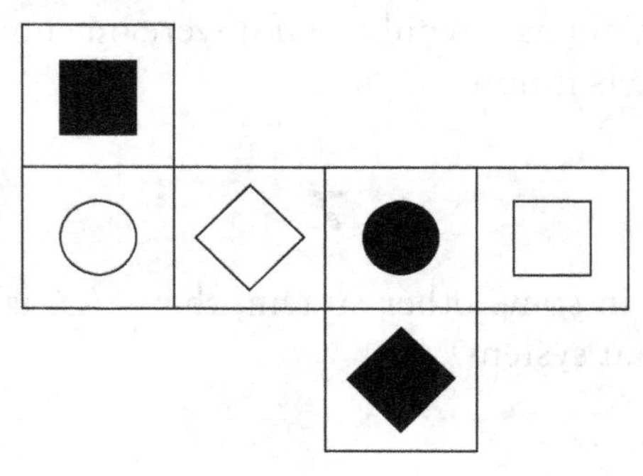

10

$2^{2^n} = 4^{4^4}$

A solution for *n* is
(a) 4
(b) 9
(c) 64
(d) 128
(e) 256

11

Here is a coin puzzle that may appear to have a counterintuitive answer, but I assure you it is correct.

I flip a penny and a dime and hide the results from you but tell you "at least one of the coins came up heads."

What is the probability that both coins came up heads?

12

The letters A U V T Y have reflection symmetry across a vertical plane. The letters H I O X have both horizontal and vertical symmetry. What capital letters have reflection symmetry across the horizontal plane only?

13

If 10! ($10 \times 9 \times 8 \times 7 \times 6 \times 5 \times 4 \times 3 \times 2 \times 1$) were to be factored into prime numbers, how many 5s would appear? What if the number were 15!, how many 5s then?

14

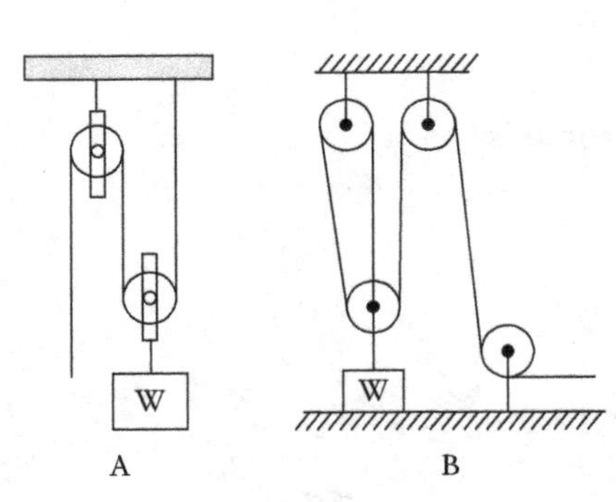

In the pulley system above:

(a) A has a mechanical advantage of 2 and B has a mechanical advantage of 3.

(b) A has a mechanical advantage of 3 and B has a mechanical advantage of 2.

(c) A has a mechanical advantage of 2 and B has a mechanical advantage of 4.

(d) Both systems have a mechanical advantage of 2.

(e) Both systems have a mechanical advantage of 3.

Which choice is the correct one?

15

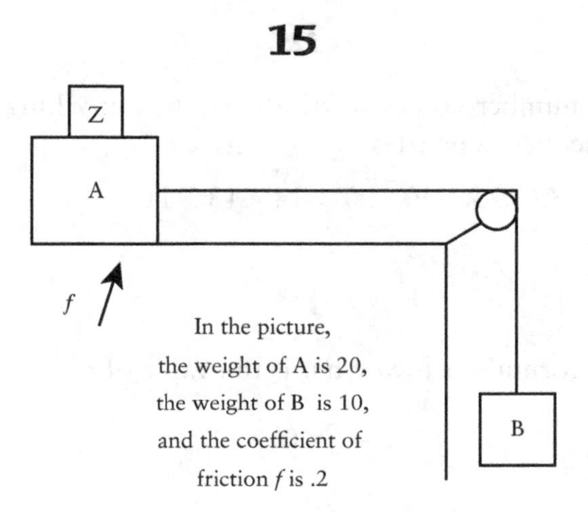

How heavy must Z be to keep A from sliding?

16

At a recent small gathering of friends, there were nine more women than men. If the ratio of women to men was 5 to 2, how many men and women were at the party?

17

I have a few objects I need to weigh with weights from 1 to 40 ounces. What are four weights I can use to be able to weigh any weight from 1 to 40 ounces (whole number weights only and including 1 and 40 ounces)? I'm trying to keep the number of weights I buy to a minimum, so what is the least number I would have to purchase of each of these four weights to accomplish my goal?

18

What number comes next? Remember, anything goes with sequence puzzles.

50 33 25 20 17 14 13 11

19

In the formula below, what is the value of g?

$$v = f \times (1 - g)^t$$

20

Given five points in space, connecting three points at a time to determine a plane (extending to infinity) what is the maximum number of lines that will result from all intersections?

21

Each figure contained within the largest square is also a square. What are the sizes of squares A, B, C, and D?

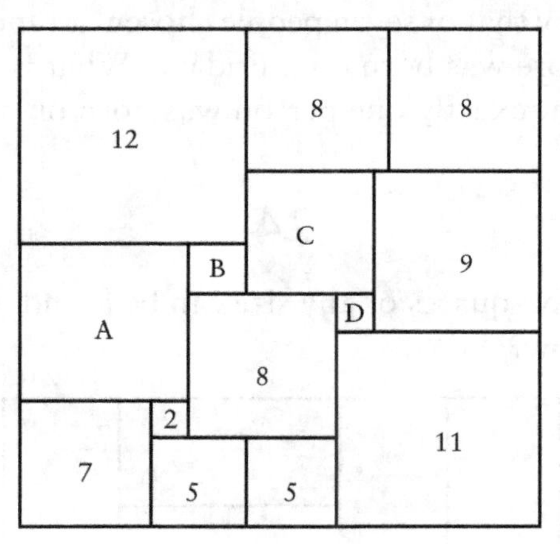

22

Tyler has a different style of eating candy. When he eats chocolates in the morning, he eats no other candy in the afternoon. If he eats chocolates in the afternoon, he has none in the morning.

In one period of time, he had no chocolate on 9 mornings and 6 afternoons, but there were 13 days when chocolate was eaten. How many total days were involved in that period?

Here are some tips that will help you solve this. There can be days when no chocolate is eaten, but any day counts when chocolate is eaten in the morning or afternoon, but not both.

23

The probability that any one person chose at random being born on a Friday is 1 in 7 or 14%. What is the probability that of seven people chosen at random, that one or more was born on a Friday? What is the probability that exactly one person was born on a Friday?

24

How many squares of any size can be found in the diagram below?

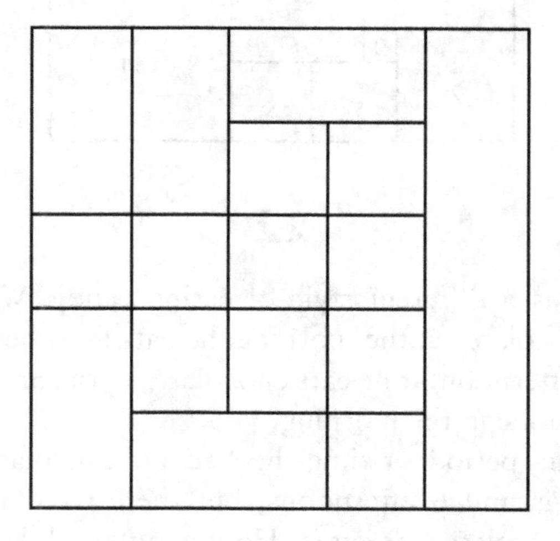

25

What number comes next in the following sequence?

0 4 –4 4 –12 4 –28 4

26

Imagine a square sheet of paper. Fold it diagonally to form a triangle, then fold it again diagonally. Snip off the three corners exactly the same size. Now open up the sheet. Which shape will it look like?

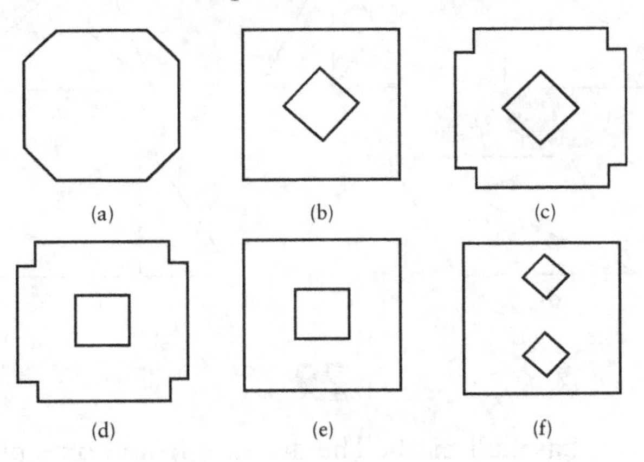

27

Given three containers with capacities 12, 7, and 6 cups, the largest is full of water; the others are empty. Show how to measure nine cups of water into the first container by pouring water from one container to another. The usual rules apply: you can pour water from one container to another until either the first is empty or the second is full. You cannot just pour water onto the ground or add water from outside the system.

28

One of the following figures lacks a basic feature the other four figures have. Which figure is the odd one out?

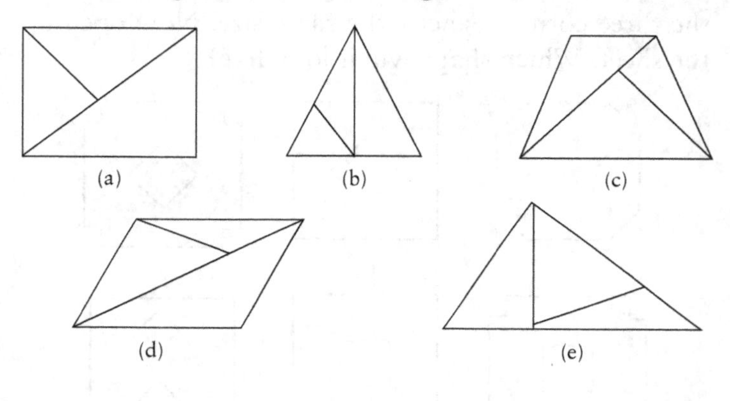

29

Try this baseball math. The distance from home plate to the pitcher's mound is 60 feet, 6 inches. A 90 mph fastball will reach the plate in 0.458 seconds.

If you moved back 90 feet from home plate, how fast would the ball have to travel to reach the plate in 0.458 seconds?

30

What numbers go in the last rectangle?

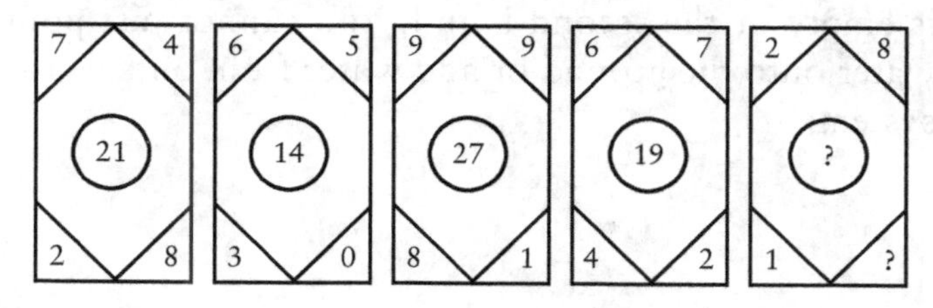

31

When the proper weights are assigned, this mobile is perfectly balanced. What are the three missing weights?

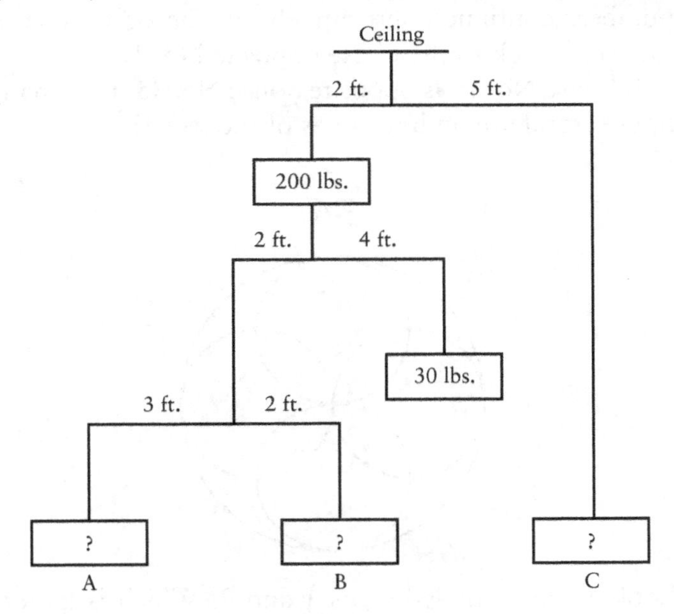

32

A group of college students took a summer job where they provided night security for a construction site. Five of the students had to be on duty every night, and the length of service for each student was to be 12 shifts only. The construction company needed security for 48 consecutive days. How many different students were needed?

33

The houses on a particular street are numbered 1, 2, 3, 4, 5, etc., up one side of the street, and then the numbers continue consecutively on the other side of the street back to the house opposite No. 1.

If house No. 12 is opposite house No. 35, how many houses total are on both sides of the street?

34

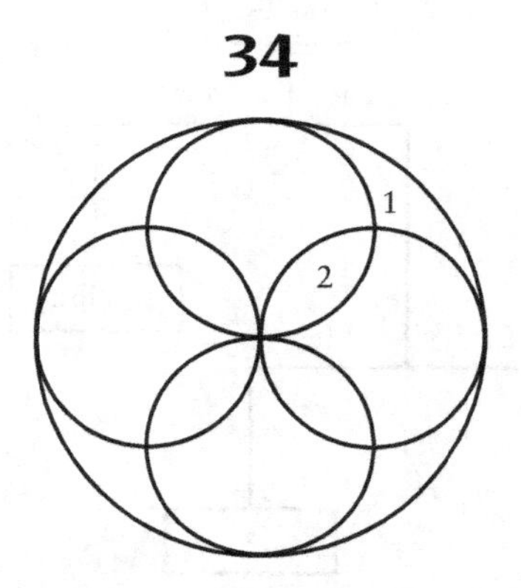

Look at the bounded areas 1 and 2. Which is larger? Are they equal? The four interior circles are equal.

35

We have two bags with two balls in each. One bag has two black balls, and the other has a black ball and a white ball. You randomly pick a bag and draw a ball, which is black.

What is the probability that the second ball you pick also will be black?

36

How many days are in 5,000,000 seconds?

37

Suppose you work at a nursery, and your boss wants you to plant rose bushes for a client. The client wants the bushes in five rows with four in each row, in a way that is the most economically feasible.

What is the minimum number of rose bushes that will be needed to accomplish this?

38

$$\sqrt{\frac{27^{4/3}}{3^2}} =$$

(a) 81
(b) 1
(c) 3
(d) 16
(e) –9

39

A chemical company is transporting 1,000 pounds of a certain chemical by truck. The chemical is 95 percent water. Part of it will evaporate during the trip. By the end of the trip, the chemical is expected to be 90 percent water. How much will the remaining chemical weigh when it reaches its destination?

40

If two typists can type two pages in two minutes, how many typists will it take to type 18 pages?

41

Using four 1s and no more than two math operations or symbols, what is the smallest positive number you can create? We'll give one answer, but there may be others.

(For this puzzle, you can't use the factorial function (!) or trig functions.)

42

Quick takes:

1) $2^{5/2} - 2^{3/2} =$
   (a) 2 (b) 2^{π} (c) $2^{3/2}$ (d) 2^{-1}

2) If the average of three numbers is w, and one of the numbers is q and another is p, what is the remaining number?
   (a) $w/q - 3 - p$ (b) $3q - p$
   (c) $3w - q - p$ (d) $w - q - p/2$

3) Evaluate $(25/64)^{3/2}$.
   (a) 25/128 (b) $5^2/8^2$ (c) 125/512 (d) 2/3

43

Illustration 1 is a square piece of paper. Illustration 2 is the piece of paper folded in half, and Illustration 3 is the piece of paper folded into fourths with the corners snipped.

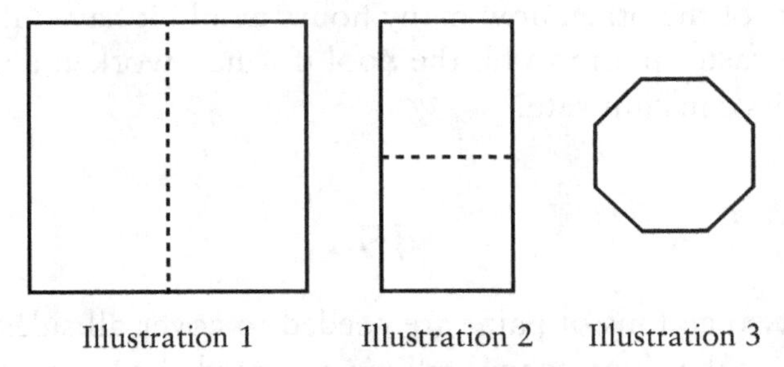

Illustration 1 Illustration 2 Illustration 3

Imagine you have snipped off the four corners as shown in Illustration 3. Now open the piece of paper.

The result will look like:

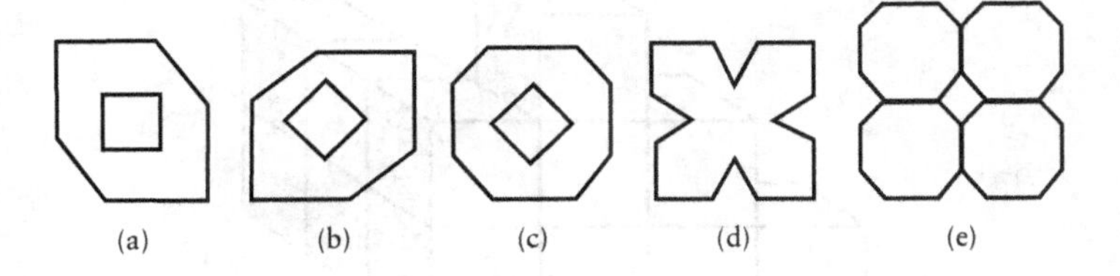

44

Two water pumps working simultaneously at their respective constant rates took exactly 4 hours to fill a swimming pool.

If the constant rate of one pump was 1.5 times the rate of the other, how many hours would it have taken the faster pump to fill the pool if it had worked alone at its constant rate?

45

If two gallons of paint are needed to cover all sides of one cube, how many gallons are needed to cover all the exposed surfaces of the figure below? Assume the bottom of the figure is exposed. Hint: There are no hidden cubes.

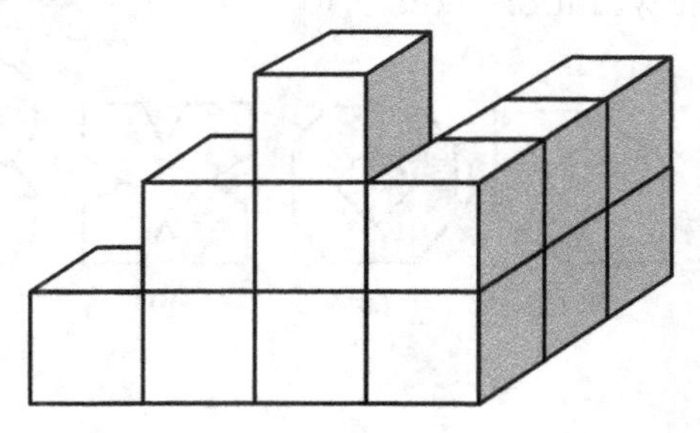

46

What number comes next?

4 7 16 43 124 367

47

Below are 40 matchsticks. What is the minimum number of matchsticks that need to be removed so there is no square of any size remaining?

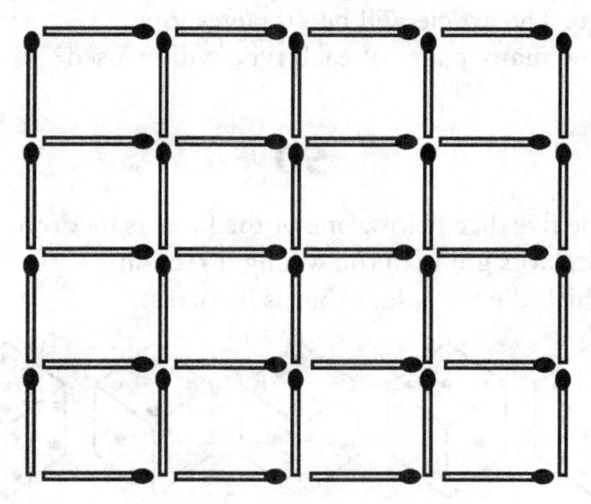

48

Two trains start from two different cities and travel toward each other at 60 mph and 50 mph respectively. At the time the trains meet, the first train has traveled 120 miles more than the second.

What was the distance between the two trains when they started?

49

A publisher is printing an article of 48,000 words. Two sizes of type will be used — one where a page consists of 900 words, and the other will have 1,500 words on a page. The article will be 40 pages long.

How many pages of each type will be used?

50

On the five dice below, one of the faces is incorrect and has the dots going in the wrong direction.

Which die has a face that is incorrect?

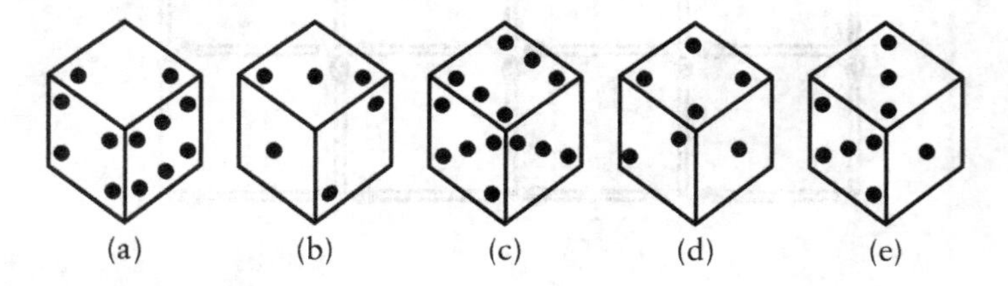

51

Below are four views of the same cube. If you were to unfold the cube, how would you fill in the diagram so that it shows the correct orientation of numbers and the letter E? We've given you a head start by placing the number 3.

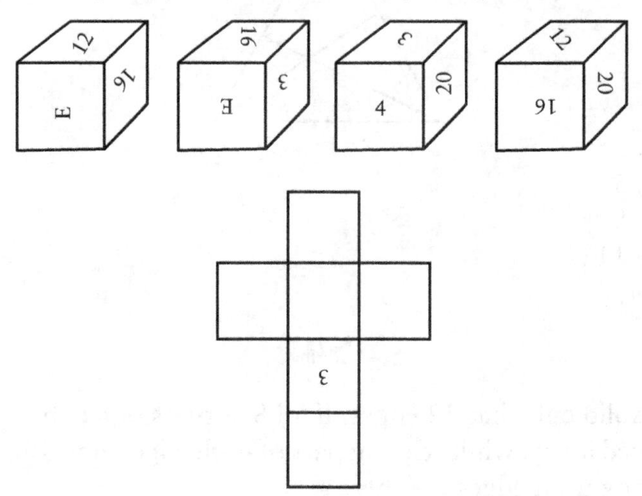

52

4/5 of a pound of cheese is balanced perfectly by 1/5 of a block of the same cheese. What is the weight of the whole block of cheese?

(a) 2 pounds
(b) 4 pounds
(c) 5 pounds
(d) 10 pounds

53

How many triangles of any size are in the figure below?

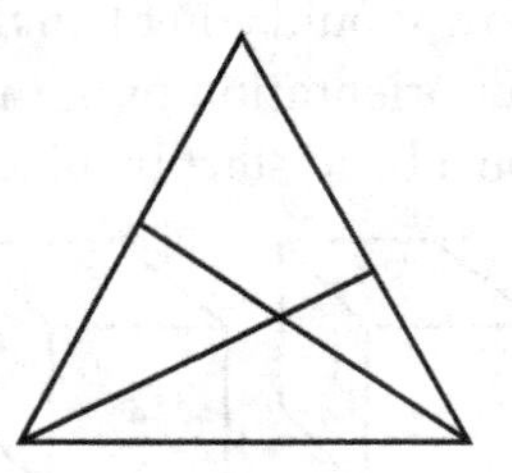

(a) 3
(b) 5
(c) 8
(d) 11

54

A solid cube has 12 edges. If all 8 corners of a cube are sliced away, while leaving part of each edge intact, how many total edges are there?

55

In each pair shown below, which is the larger, or are they the same?

(a) $\dfrac{1}{\dfrac{\sqrt{.1}}{\sqrt{.01}}}$ or $\dfrac{1}{\dfrac{\sqrt{.01}}{\sqrt{.1}}}$

(b) 3^{18} or 9^{9}

(c) $2^{23} + 2^{23}$ or 2^{24}

56

$2^{32} + 1$ is exactly divisible by a whole number. Which of the following numbers is exactly divisible by this same number?

(a) $2^{16} + 1$
(b) $2^{16} - 1$
(c) 7×2^{23}
(d) $2^{96} + 1$

57

On the face of a digital clock, like the one below, there are 28 line segments to represent the time — of course, not all are used at the same time. One of these segments is used more than any other one. Can you determine which one it is?

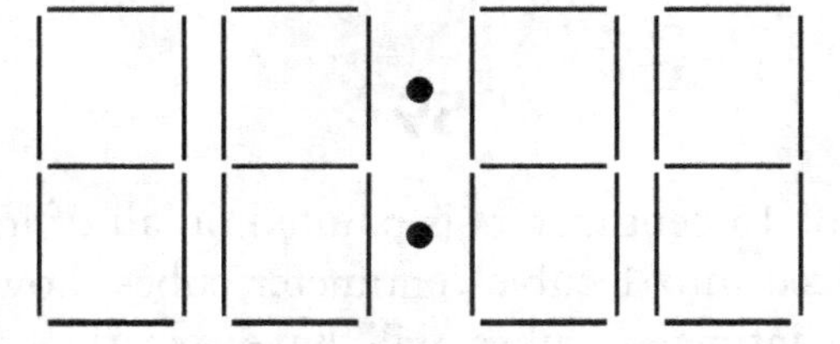

58

How many cubes are in the configuration below? All rows and columns run to completion unless you actually see them end.

59

A cube of 15 centimeters is painted on all of its sides. If it's sliced into 1-cubic-centimeter cubes, how many 1-cubic-centimeter cubes will have exactly one side painted?

(a) 125
(b) 528
(c) 1001
(d) 1014
(e) 1375

60

There are five Leeps in a Grat, seven Grats in a Bliz, and three Blizzes in a Zank.

What is the number of Leeps in a Zank divided by the number of Grats in a Zank?

61

Nine identical sheets of paper are used to create the design below. If D was placed first and F placed seventh, in what order would G be placed?

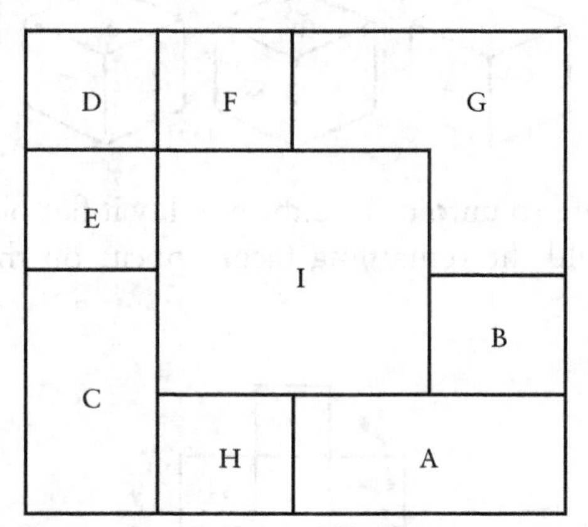

62

What is x if

$4^x \cdot 16^{-x} \cdot 64^x = \sqrt[4]{4}$?

63

See if you can figure this one out.

100 students are majoring in math, business, or both. 72 percent of the students are business majors, and 58 percent are math majors.

How many students are majoring in both?

64

Below are three views of the same cube.

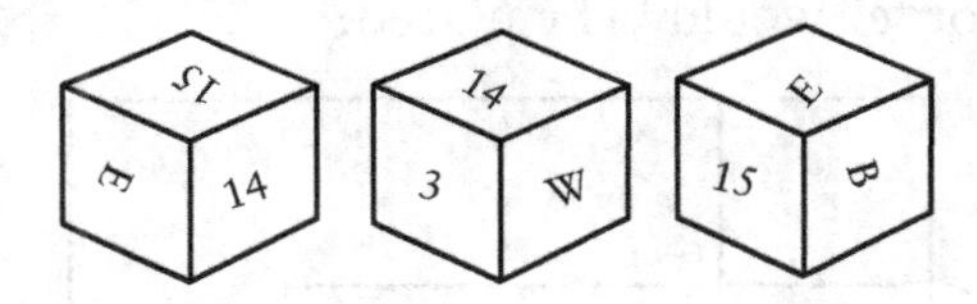

If you were to unfold the cube and lay it flat on a table, how would the remaining faces appear on the layout below?

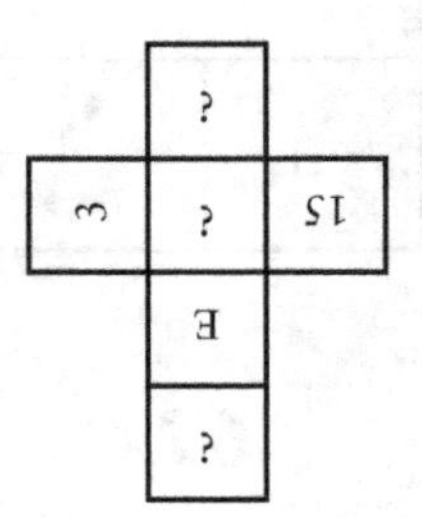

65

Here's a fun puzzle you can do with pennies (or any coins for that matter).

A total of 13 pennies are put into three piles, so that each pile has a different number of pennies.

What is the smallest number of pennies that could be in the largest pile?

66

A message was found in a strange code that read FLANG BRONO HASSI. It was determined that this meant WE COME IN PEACE. Then a message was found that read MANSO NASI FLANG, and this meant PEACE, LOVE, AND FREEDOM. Finally, BRONO BRANGO VANX was determined to mean IN THE LAST MINUTE. What does HASSI mean?

67

How can you create an answer of 64,000 by multiplying two numbers, neither of which contains a zero?

68

At a certain school, the ratio of boys to girls is 2 to 7. If eight more boys attend the school this year, the new ratio of boys to girls will be 1 to 3.

How many boys currently attend the school?

69

On a standard die, the opposite faces total 7. In our version, none of the opposite faces total 7. Below is our die from four different perspectives.

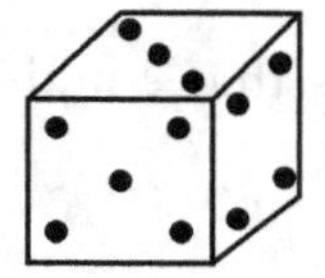
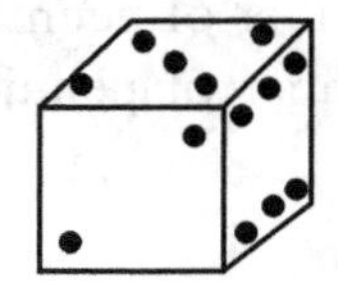
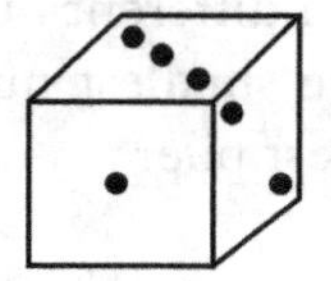
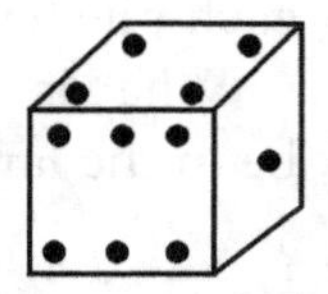

How many dots are on the side opposite the side with three dots?

70

Using 10, 20, 30, 40, 50, and 60 (50 is already given), can you assign the weights to make this mobile perfectly balanced on all levels? Each weight may be used once only.

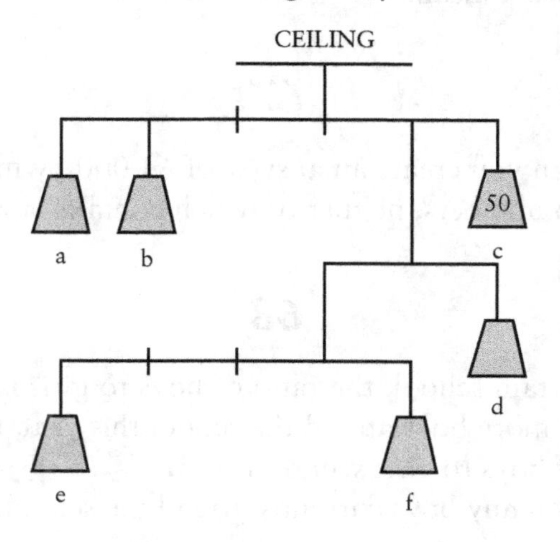

71

Molly drives 42 miles to work. Due to construction, she can average only 10 mph on the first 21 miles of the trip.

How fast must Molly go on the second half of the trip to average 19 mph for the whole trip?

72

The number $\frac{2^{58}+1}{5}$

(a) is an odd integer
(b) has a remainder of 2
(c) is an even integer
(d) has a remainder of 4

73

If today is Monday, and I told you I would meet you for dinner three days after two days before the day before tomorrow, when would I meet you?

74

Divide 50 into two parts so that the sum of their reciprocals is 1/12.

What are the two numbers?

75

The diagram shows three normal dice. How many dots are on the side of the last die that is facing the middle die?

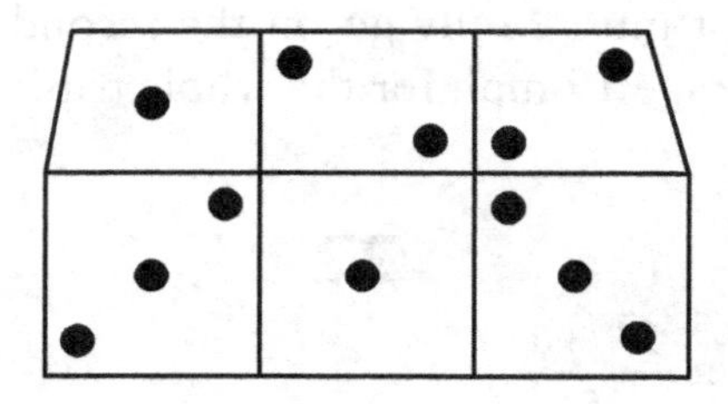

76

One hundred dental specialists are attending a medical convention. Each specialist is either an orthodontist or an oral surgeon. At least one is an orthodontist. Given any two of the specialists, at least one is an oral surgeon.

How many are orthodontists and how many are oral surgeons?

77

There are two ls between 1 and 10 (inclusive) and 21 between 1 and 100 (inclusive).

How many 1s are between 1 and 1,000,000 (inclusive)?

78

Bella is doing a lab experiment and just realized she has put a certain solution in the wrong cylindrical jar. She needs another cylindrical jar with a 30% larger diameter and the same volume as the jar she's currently using.

If the diameter of the new jar is increased by 30% without the volume changing, by what percent must the height be decreased? Remember, the volume of a cylinder is $V = \pi R^2 \cdot h$

79

What is the largest sum of money that cannot be reached using just two chips if one is valued at $10 and the other at $11?

Of course, you can use as many chips of each as you want.

80

What is the probability of rolling a six when a die is tossed twice?

(a) 1/2
(b) 1/36
(c) 11/36
(d) 1/12
(e) 7/36

81

Walking at 3/4 of her usual speed, Molly arrives at her destination 2 1/2 hours late.

What is her usual time for walking this distance?

82

The following puzzle features analytical reasoning. See if you can determine the relationships between the symbols and letter combinations to find solutions to the two unknowns.

***= BOP

*
*= SOZ

o o= BIZ

o
o= SIP
o

**= ?

SIZSOP = ?

83

What is the probability of throwing six dice all at once and having them come up 1, 2, 3, 4, 5, and 6?

(a) 1 out of 16
(b) 1 out of 65
(c) 1 out of 125
(d) 1 out of 955
(e) 1 out of 10, 275

84

There's an old puzzle that asks the question: "If a hen and a half lays an egg and a half in a day and a half, how many hens will it take to lay six eggs in six days?" The answer is 1.5 hens. Now here's the new twist: At this rate, how many eggs will one hen lay in one day?

85

Below is a magic square where the sum of the numbers in each row, column and diagonal is the same. What number replaces the question mark?

	32	
8		40
36	?	20

86

The first three terms of an arithmetic sequence are $2x - 2$, $5x - 10$, and $3x + 2$.

What are the three numbers?

87

In how many different ways can four identical cubes be placed face to face?

88

The ratio of the number of juniors to seniors in a certain school club was 3:8.

There were 40 more seniors than juniors before half the juniors and some of the seniors left to join another activity. Then the ratio of the number of juniors to seniors became 3:4.

How many of the seniors left the club to join the activity?

89

Each face of a cube will be painted either red or blue. How many different possible paint combinations are there, from all blue faces to all red faces, and every combination in between?

(a) 36
(b) 64
(c) 101
(d) 128
(e) 256

90

If Linda walks to work and rides back home, it takes her 3 hours total.

When she rides both ways, it takes her 1 hour and 15 minutes.

How long would it take her to make the round trip by walking?

91

Below are four statements. Your job is to determine how many, if any, could be true.

(a) Only one of these statements I'm writing is false.
(b) Only two of these statements I'm writing are false.
(c) Only three of these statements I'm writing are false.
(d) All four of these statements I'm writing are false.

92

If each of the numbers in the expression $wx^2y^4z^3$ is decreased by 25 percent, by what percentage is the value of the expression decreased?

93

How many cubes are in this stack? All rows and columns run to completion unless you actually see them end.

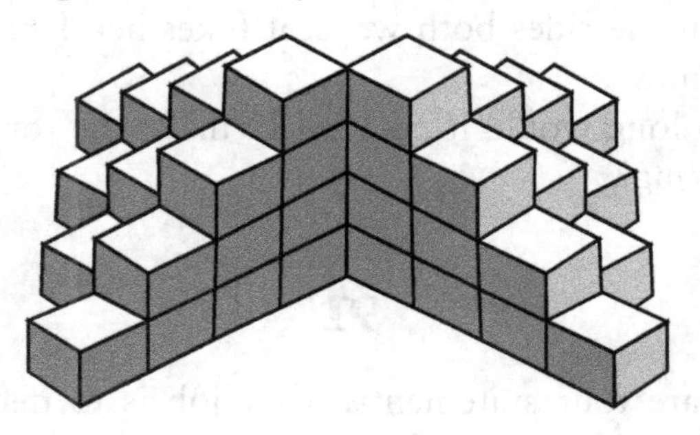

94

How many squares of any size are in the figure below?

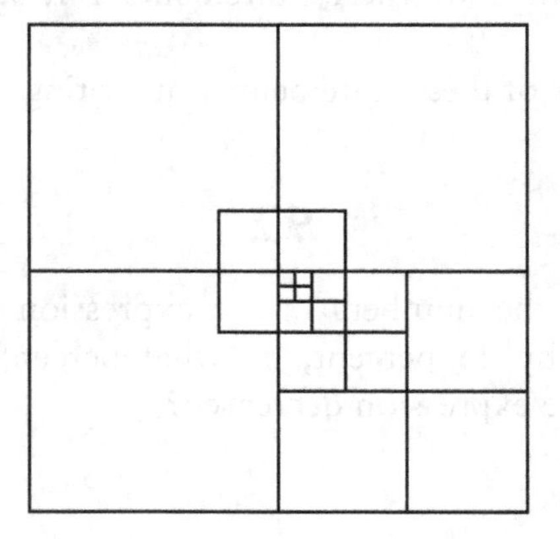

95

Karen walked to school at an average of 3 miles an hour, and she jogged back home along the same route at 4 miles per hour. If the total time going and coming back was exactly one hour, what was the total number of miles in the round trip?

96

A software engineer is going to write a special program for a top-secret government project. The software will print out a certain number of pages based on the total number of digits used for the pages. As an example, let's say the software engineer was told she could use only 35 total digits in the page numbers (starting at page 1 and continuing with the counting numbers). That means if we count all the digits in those page numbers we would have 22 pages. Pages 1–9 and two digits each from page 10–22 equal 35 total digits. When she received a final go-ahead with all the secret instructions, she was to use 1770 digits. How many pages were in the document?

(The count begins with 1, not 0, but the zeroes in 10, 100, 1000, etc., are counted as digits.)

97

A container holds 1500 oz. of a 40% salt solution.

How much pure water must be added to make the solution 10% salt?

98

What percent of 15 is 15 percent of 1/15?

99

Eleven triangles result when connecting the diagonals of a regular pentagon. What is the maximum number of triangles of any size if you include all the triangles that can be further subdivided? An example would be triangle ABD.

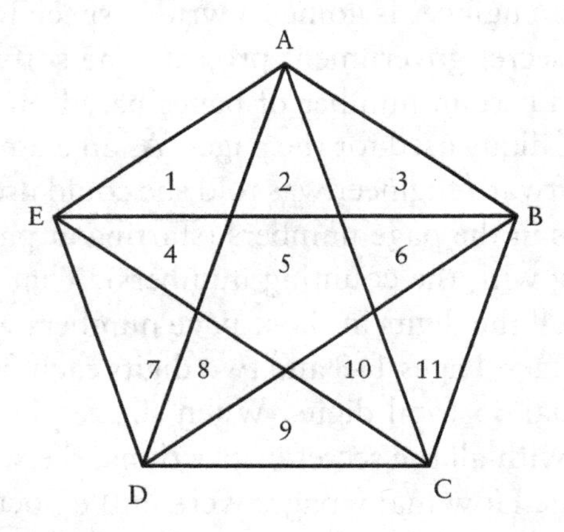

100

What is the tenth digit to the right of the decimal point in the decimal expansion of $(1/5)^{10}$?

(a) 0
(b) 2
(c) 4
(d) 6
(e) 8

101

If 3 hours ago it was 1/2 as long after
1 o'clock in the afternoon as it was before
1 o'clock in the morning, what time is it now?

102

$\sqrt{7}$ percent of $4\sqrt{7}$ is
(a) 7/25
(b) 12
(c) 1/3
(d) impossible to evaluate

103

In how many distinct ways can the digits in the number 99986 be arranged?

104

The dreaded cube-eaters from the fourth dimension descend upon a stack of 27 identical sugar cubes. Cube-eaters can eat only to the center of a cube. When they reach the center, they always make a 90-degree turn and proceed to the next cube. They never re-enter a cube. If a cube-eater enters at location A, what is the minimum number of cubes it will eat through to reach the cube at location B?

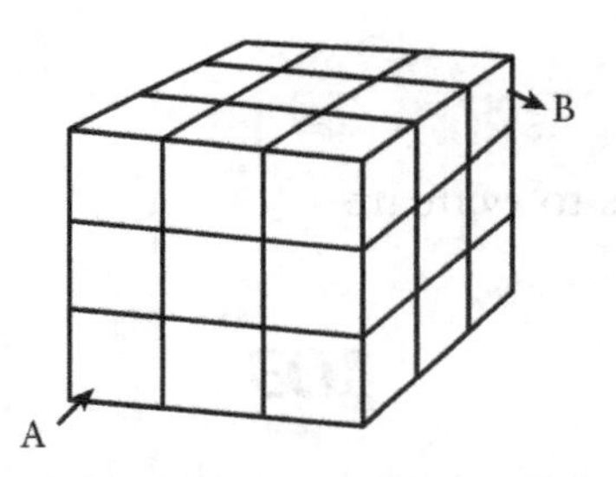

105

Martha, Mike, and Minnie are going to play a series of one-set tennis single matches against each other. The winner of each set stays on the court and faces the player who has been idle. The loser of each set sits out the next match. At the end of the week, Martha played 15 sets, Mike played 14 sets, and Minnie played 9 sets. Who played the 11th set?

106

A softball infield consists of a first baseman, second baseman, shortstop, and third baseman. They are Brian, Rich, Kendall, and Steve, but not necessarily in the order listed above.

(a) Brian is younger than Kendall.
(b) The first baseman is not related to any of the other infielders.
(c) The second baseman and shortstop are brothers.
(d) Kendall is Rich's nephew.
(e) The second baseman is not the third baseman's uncle. And the third baseman is not the shortstop's uncle.
(f) What position does each man play, and how are they related?

107

How many squares of any size are there in the figure below?

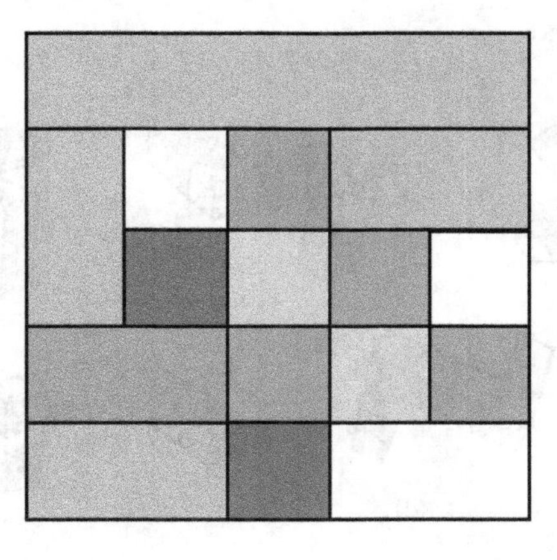

108

Cashews sell for $8 a pound and peanuts sell for $3 a pound. A store wants to make 20 pounds of mixed nuts that will sell for $6 a pound.

How many pounds of cashews and how many pounds of peanuts should be used?

109

A bicycle is three times as old as its tires were when the bicycle was as old as the tires are now.

What is the ratio of the tires' current age to the bicycle's current age?

110

Are the following five orientations of cubes the same or different? If you think there is a difference in the way one or more of the choices is oriented, can you identify which ones?

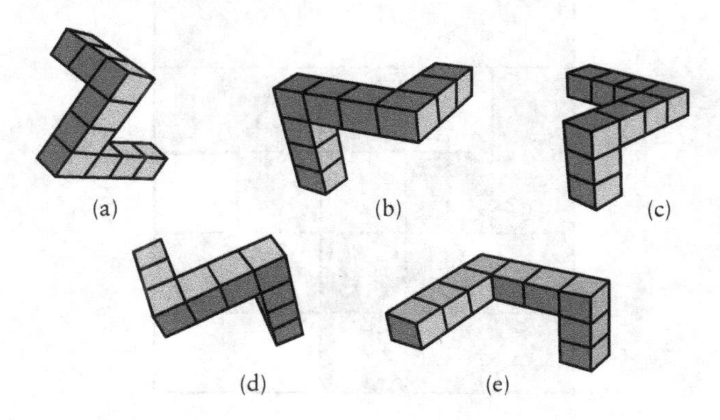

111

There is an old puzzle that asks the following:

At one point, a remote island's population of chameleons was divided as follows:

13 red chameleons
15 green chameleons
17 blue chameleons

Each time two different-colored chameleons would meet, they would change their color to the third one. (E.g., if green meets red, they both change their colors to blue.) It is not possible for all the chameleons to become the same color. What are two examples (each with different sets of numbers and colors) that will work?

112

Using the numbers 5, 6, 7, and 8 once and only once, what is the largest number that can be created by "stacking" the numbers in exponential form?
What are the second and third largest?

Examples: 6^{875} $5^{76^{8}}$ $87^{6^{5}}$

You may stack the numbers any way you choose in any combination to achieve the largest number.

113

How many different sides or faces does this solid have? Block A is longer than Block B from left to right.

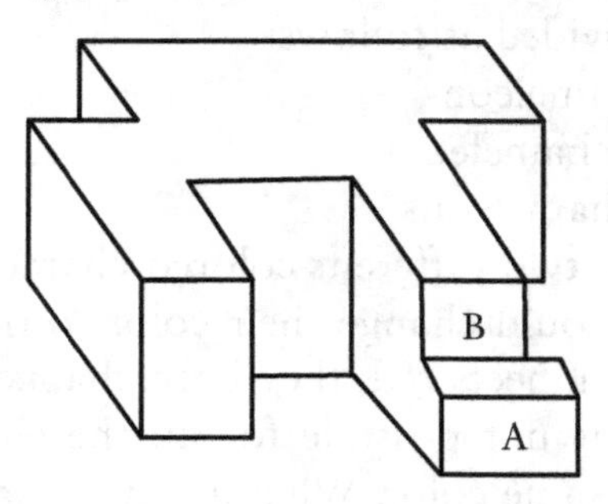

114

Molly drives 42 miles to work. Because of construction, she can average only 12 miles per hour for the first 21 miles of the drive.

How fast must she go on the second half of her trip to average 20 miles per hour for the entire trip?

115

A microscopic slide shows 7,500 bacteria dying at a rate of 150 per hour. Another slide shows 4,500 bacteria increasing at a rate of 50 per hour. In how many hours will the bacterial count on both slides be the same?

116

If .1 percent of x is equal to 10 percent of y, then 10 times y is what percent of x?

117

An escalator is moving downward at a constant speed. Andrea walks down and takes 50 steps to reach the bottom. Marty runs down and takes 90 steps in the same time as Andrea takes 10 steps.

How many steps are visible when the escalator is not moving?

118

What is the value of $\sqrt{.2^{-2}}$?

(a) 5
(b) 25
(c) 10
(d) .04
(e) −.04

119

Below is a list of scores from a fictitious college football season. Based on the given scores only, if Central Michigan were to play Michigan during this season, who would win and by how much?

Michigan 27, Nebraska 13
Nebraska 14, Oklahoma 10
Northern Iowa 24, Central Michigan 21
Oklahoma 10, Baylor 7
Central Michigan 28, Minnesota 24
Minnesota 35, Baylor 3

120

At a luncheon for physicians, all but 40 were neurosurgeons. All but 50 were pediatricians, and all but 60 were cardiologists.

How many physicians were there in all?

121

See if you can do the math on this one.

If a team wins 40 percent of its games in the first one-fifth of the season, what percentage of the remaining games must it win to finish the season having won 70 percent of the total games played?

122

Molly sold 6,000 shirts in one year. This was 140 percent more than she had sold the year before.

How many shirts did Molly sell the year before?

123

What is the inverse of

$$y = \frac{x-2}{x+3}?$$

124

You have two candles. Both candles will stay lit for 60 minutes. Your task is to find a way to measure 45 minutes by burning one or both. You cannot cut the candles in half.

How can this be accomplished?

125

In a 7-by-7-by-7 stack of cubes that are all the same size, how many of the individual cubes are completely hidden from view?

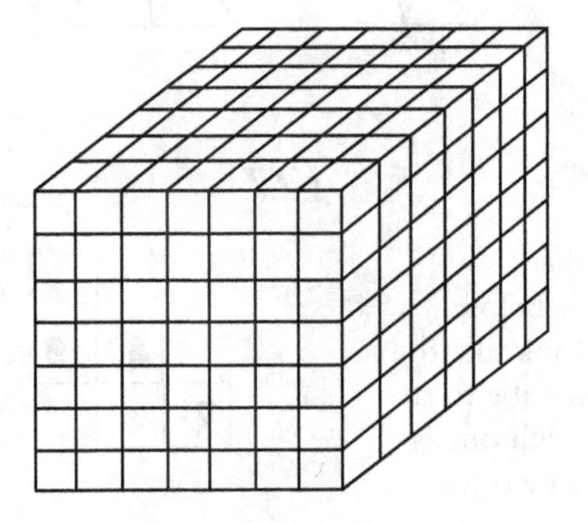

126

One of the figures below is different from the rest based on a simple design reason. Which is the odd one out?

Hint: It does not have to do with symmetry, rotation or 90° angles.

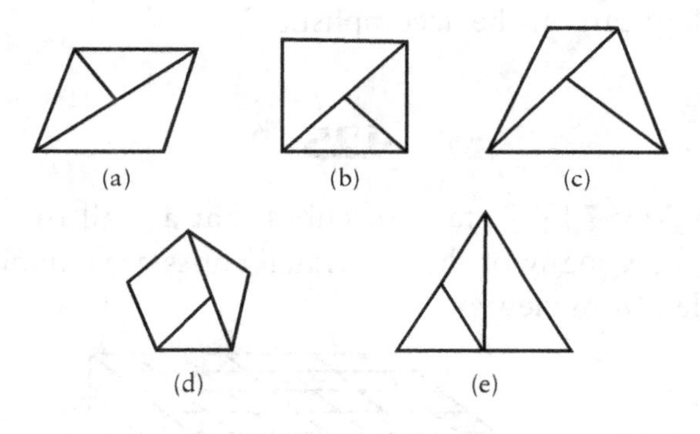

127

One of the five cubes below doesn't look like the unfolded cube to the right. Which one is the odd one out?

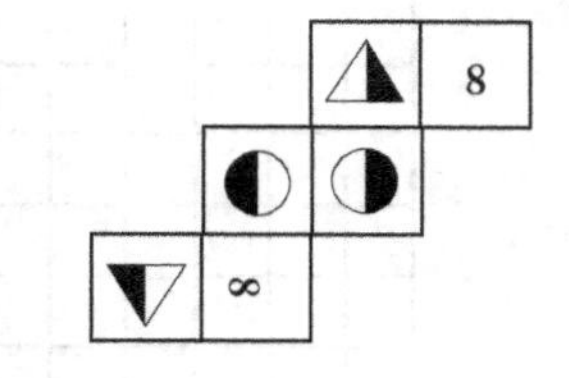

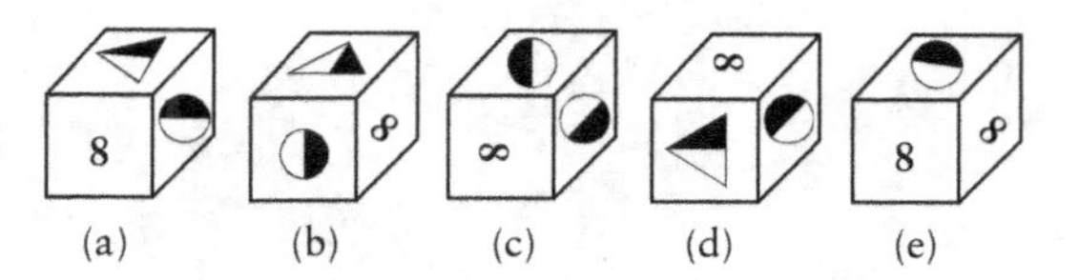

128

Eddie Fontani, head of the local racetrack, was shot and killed gangland-style. The police brought in five subjects for questioning. Each made three statements but the police knew that for each person, two of the statements were true and one was false. Who killed Fontani?

Louie:	I didn't kill Fontani. I have never owned a .38. Sawed-Off Pete did it.
Robbie:	I didn't shoot Fontani. I never owned a pistol. The others are all passing the blame around.
Doogie:	I am innocent. I never saw Benny before. Sawed-Off Pete is guilty.
Sawed-Off Pete:	I didn't do nuthin' to nobody. Benny is the guilty one. Louie did not tell the truth when he said I did it.
Benny:	I had nuttin' to do wit' Fontani's murder. Robbie is the one who did it. Doogie and I are old pals.

129

Jim and Sally live 100 miles apart. Each drives toward the other's house along a straight road that connects them. Jim drives at 65 mph and Sally drives at 55 mph.

If they leave their homes at the same time, how long will it take them to meet?

130

What fraction in $Base_{10}$ is equal to .132 in $Base_4$?

131

Below is the beginning of a sequence that continues to infinity.

0 2 6 12 20 30 42

What is the 100th number of this sequence?

132

Below is an unfolded cube with the numbers 1, 2, and 4 on three of its faces. Can you place the numbers 3, 5, and 6 so that when the cube is reformed, the opposite faces will total seven?

Note: For this puzzle the orientation of the numbers is not necessary. Getting the correct number on each face is all that counts.

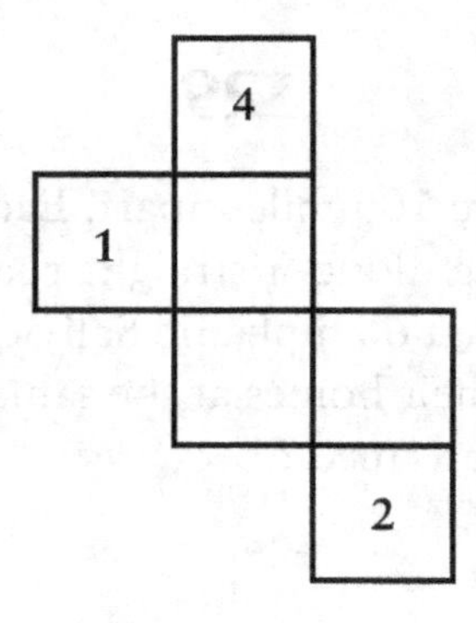

133

Can you create just two squares from the figure below by moving only one line and without leaving any leftover lines?

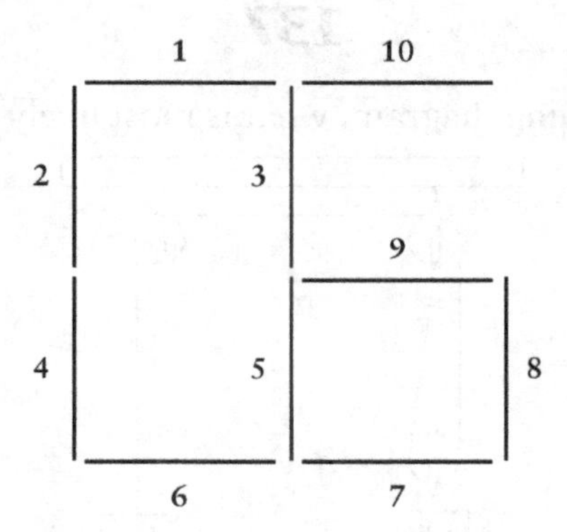

134

Here is a riddle...

30 heads, 100 feet.
Both birds and bulls are there to meet.
But how many of each are there now?
Can you show me the way?
Oh me! Oh wow!

135

Imagine a circle with diameter AB. Choose a point outside the circle; we'll call it point P. Using a straightedge only, construct a perpendicular line to diameter AB.

No protractors or compasses, please.

136

If p and q are prime numbers, how many divisors does the product p^4q^7 have?

137

In the following diagram, what is most likely to happen?

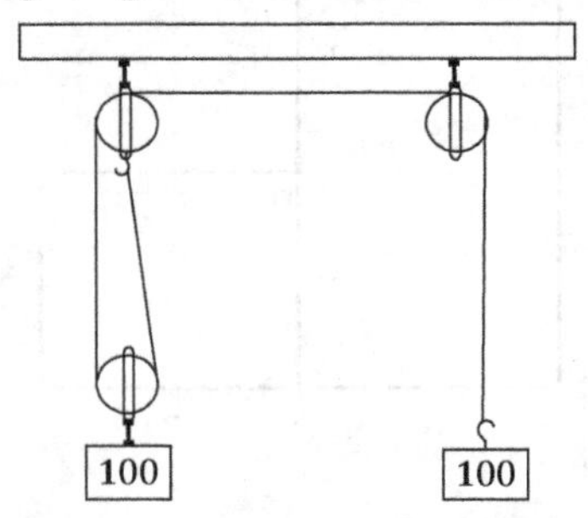

(a) The right-hand weight will travel downward.
(b) The left-hand weight will travel downward.
(c) The weights will not move.
(d) Unable to determine this from the information provided.

138

Molly works out in a gym on the first floor of a large building almost every day and she walks up an escalator that connects the ground and the first floor. If she walks up the escalator step-by-step, it takes her 20 steps to reach the first floor. One day, she doubles her stride length and walks up climbing two steps at a time. It takes her 16 steps to reach the first floor.

If the escalator stood still, how many steps would there be that would be visible on sight?

139

You have two pieces of string of varying thickness that each burn for 40 seconds.

You want to measure 30 seconds. You can't cut either string in half because the fuses are different thicknesses, and you can't be sure how long each will burn.

How can you accomplish this without bending, folding, or cutting the strings?

140

Nine identical sheets of overlapping paper are used to create the design below. If D is placed first, F is placed seventh, and I is placed last, in what place would you find B? Where would you find H?

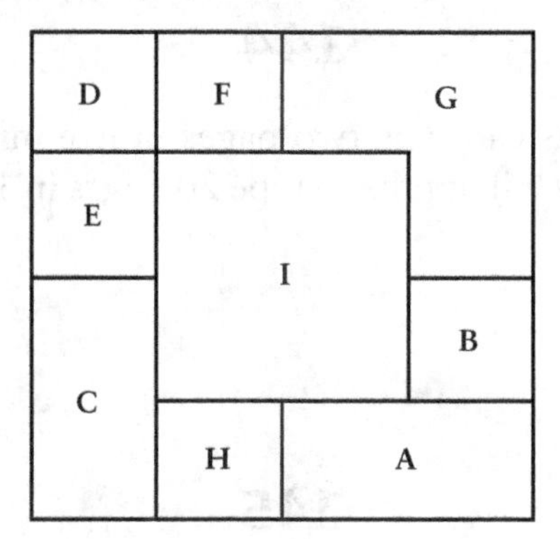

141

How many times do the hour and minute hands of a clock form a right angle in 24 hours?

142

Twelve people start a new software company. If there had been four more initial investors, the initial expense for each person would have been $2,000 less.

What was the initial cost per person?

143

Find the sum of the values of w, x, y, and z, using the problem below to determine the value of each letter.

$$\begin{array}{r} 5w \\ +xy \\ \hline z43 \end{array}$$

144

If two typists can type two pages in five minutes, how many typists will it take to type 20 pages in 10 minutes?

(a) 5
(b) 10
(c) 15
(d) 20

145

Bob drives his scooter 2.5 times as fast as Bibi runs. Together they cover a total of 42 miles in one hour.

What is their combined distance if Bob drives his scooter for 1/2 hour and Bibi runs for 1 1/2 hours?

146

If the stack of cubes below was originally 3 × 3 × 3, which of the pieces below is the missing piece of the cube? All rows and columns run to completion unless you actually see them end. The missing piece is to be inserted upside down to complete the cube.

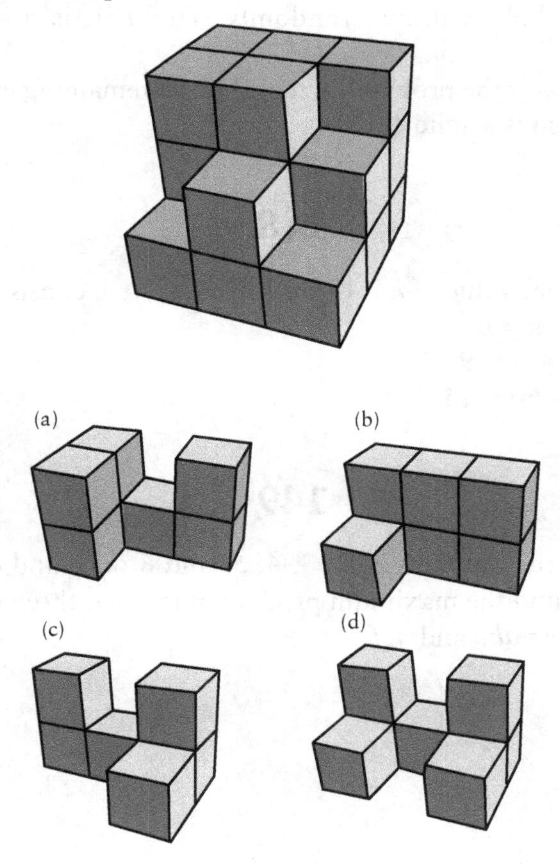

147

A bag contains one solid black pool ball that cannot be seen. A second pool ball is randomly chosen from another bag in such a way that it is equally likely to be a solid color or a striped ball. That ball is then added to the original bag. The original bag is shaken, and a ball is drawn randomly. This ball is a solid color.

What is the probability that the ball remaining in the bag also is a solid color?

148

For what value of k is the following system consistent?
(a) $p + q = 6$
(b) $kp + q = 9$
(c) $p + kq = 15$

149

Using the numbers 1, 2, 3, 4, 5, and 6 once and only once, find the maximum product of the two three-digit numbers *abc* and *def*.

150

If you were able to start on any edge of a regular cube and trace a path from edge to edge without going back over any edge, you would find that the maximum number of edges you could trace would be 9.

What is the maximum number of edges that can be traced on a regular octahedron?

151

A man can hit the bullseye of a target one out of three times.

If he shoots four shots in succession, what is the probability of successfully hitting the target at least once?

152

You may be familiar with the Fibonacci sequence:

0, 1, 1, 2, 3, 5, 8, 13, 21 ....

Each number is the sum of the preceding two consecutive numbers.

What are the six consecutive numbers to the left of 0 in the Fibonacci sequence?

153

If the stack of cubes below was originally $4 \times 4 \times 4$, which figure represents the missing piece from the cube? All rows and columns run to completion unless you actually see them end.

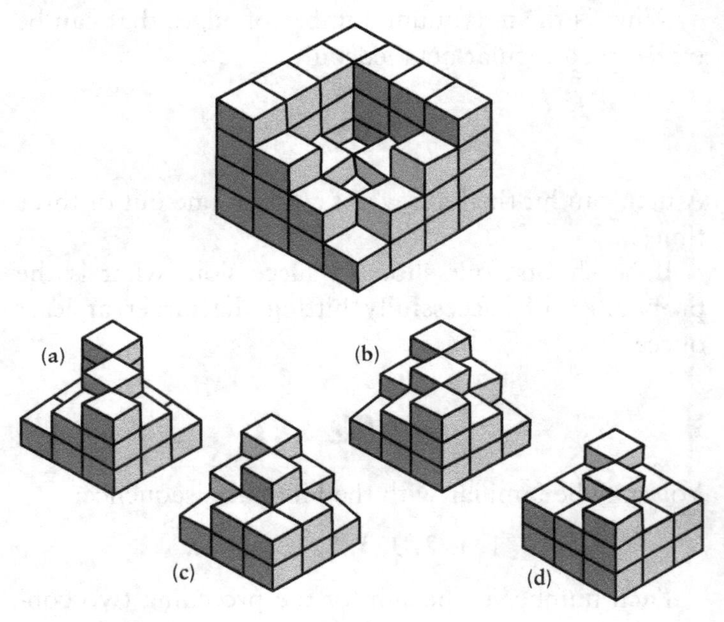

154

Let $a + b + c = 4$ and $d + e = 5$. What is $ad + be + cd + ae + bd + ce$?

155

Consider six electrical lights. In how many ways can they be networked so that each light is directly connected to two other lights, and all the lights are connected directly or indirectly?

156

I purchased 100 pounds of potatoes, which I knew to be exactly 99 percent water. After drying in the sun for a week, they were only 98 percent water.

What is the total weight of the potatoes after drying?

157

Each symbol in the puzzle square below represents a whole number. The numbers outside the square are the sums of each row and column. Can you determine the value of each symbol?

*	◆	○	*	30
○	◆	○	◆	40
○	◆	*	○	34
*	*	◆	*	26
28	38	34	30	

158

There are eleven frogs, each on a different lily pad. No two distances between any pair of lily pads are the same. Each frog is going to jump to the nearest pad. When this happens, will there be any pad that doesn't have a frog on it?

159

3 7 8 2
9 9 0 2
4 5 5 6
1 8 9 2
2 6 3 9
5 4 8 3
? ? ? ?

The number that would be most logical to replace the question marks is

(a) 6 1 8 5
(b) 7 9 9 7
(c) 8 8 3 3
(d) 9 0 6 4

160

Here is a difficult math brainteaser:

If our number system were based on 12 instead of 10, then we would need two new symbols, one for 10 and one for 11. 12 in $Base_{10}$ would become 10 in $Base_{12}$. Let's say the new symbol for our 10 is *x* and the new symbol for our 11 is *y*.

How would you write our 1000 ($Base_{10}$) in $Base_{12}$?

161

Your chemistry teacher has just asked you to convert temperatures from one system of measurement to another. You are told 14° in the first system is equal to 36° in the second system. Likewise, 133° is equal to 87° in the second.

What is the formula for converting one system to another? At what temperature will both thermometers read the same?

162

Counting by 3s and starting with 1 results in the sequence 1, 4, 7, 10, 13, 16...

What is the 374th number in this sequence?

163

$2^{2^{-2}} = ?$

(a) –8.314
(b) 8
(c) –16
(d) 1.189
(e) π

164

Your school is holding a raffle. There are 600 raffle tickets in a bowl, numbered 1 through 600. For a small fee, each participant can reach in and take out one ticket to win a door prize. One of the students in the math club asks her father, "What is the maximum number of tickets that must be drawn to ensure that one of the numbers on the ticket is exactly twice as large as a previously chosen number?" How many would have to be drawn?

165

If the stack of cubes below was originally 3 × 3 × 3, which one of the following figures is the missing piece from the broken stack of cubes?

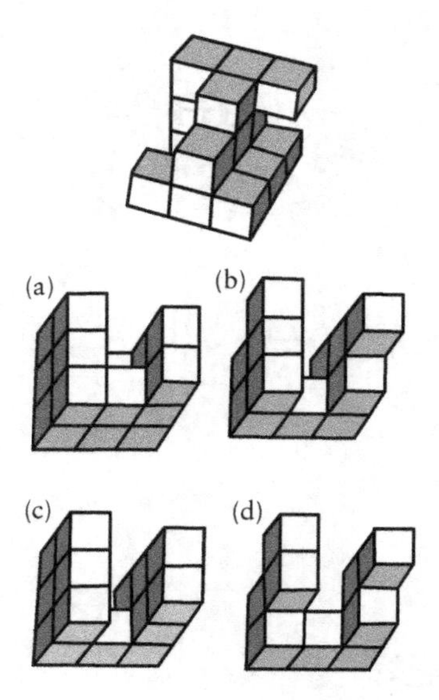

166

As an experiment in a physics class, a perfectly circular ring with a diameter of 20 inches is set in motion around another ring with a diameter of 60 inches. The smaller ring adheres to the surface of the larger ring in such a manner that it rotates around the larger ring with no slippage.

How many revolutions will the smaller ring make around the larger ring if it starts at the very top, makes one revolution, then returns to the point from which it started?

167

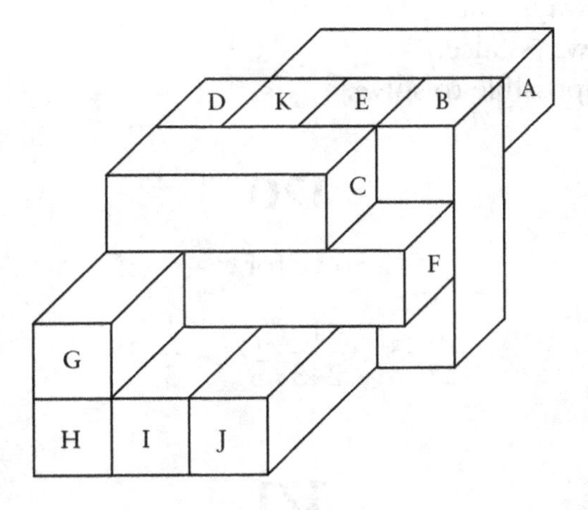

In the above stack of blocks, there are eleven total blocks, and they are the same size. How many times does each block touch the faces or sides of adjacent blocks? Edges do not count. For example, D touches C, G, and H, so it has three touches. Another example would be G making four touches — C, D, F, and H. Fill in the touches for each remaining letter.

168

The fraction 1/9 is equal to the repeating decimal .1111...

What is 1/9 in Base_{12} decimal form?

169

If $x < 1$ is the following true or false?

$$x\,|x|^3 < |x|^x$$

$|x|$ means the “absolute value” of x.

(a) Sometimes true.
(b) Always true.
(c) Always false.
(d) Impossible to solve.

170

Solve for x:

$$\frac{1}{27} \cdot 3^{59} \cdot \frac{1}{243} \cdot 27^x = \frac{1}{3} \cdot 3^x$$

171

If $a = x$ percent of y and $b = y$ percent of x, then:

(a) a is smaller than b.
(b) a is greater than b.
(c) a is equal to b.
(d) No relationship can be determined.

172

Four normal dice are side by side. How many dots are on the face with the question mark?

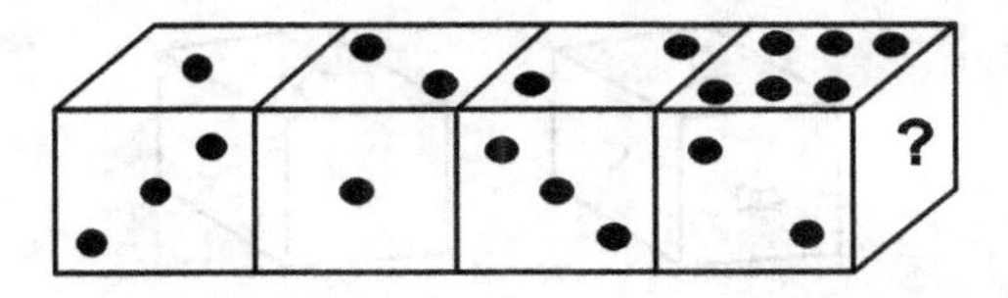

173

A ladder was standing perfectly upright against a wall. Suddenly, the foot of the ladder slid away from the wall and came to a stop 15 feet from the wall. The top of the ladder had moved only 1/5 of the ladder's length before it came to rest firmly on a windowsill.

Is there enough information to calculate the exact length of the ladder? If so, what is it?

174

In the land of Q.E.D., the number system resembles ours in the symbols used, but the results of its math operations are different. For instance, $4 \times 2 = 12$, $5 \times 3 = 23$ and $2 \times 3 = 10$.

What does $2 \times 3 \times 4$ equal in its number system?

175

Below are three different views of the same cube. What letter is on the face opposite H?

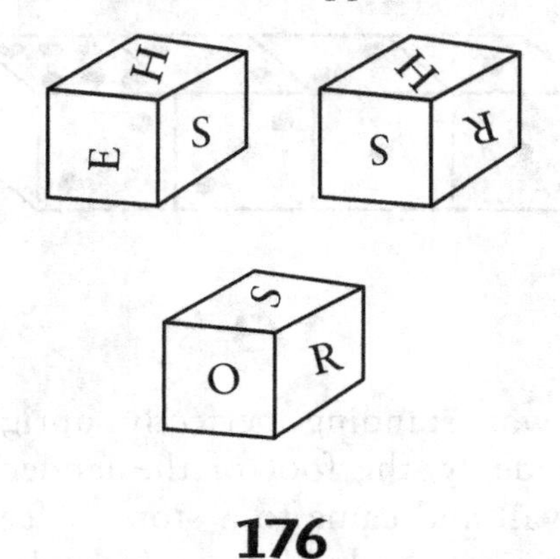

176

How many different lines are there?

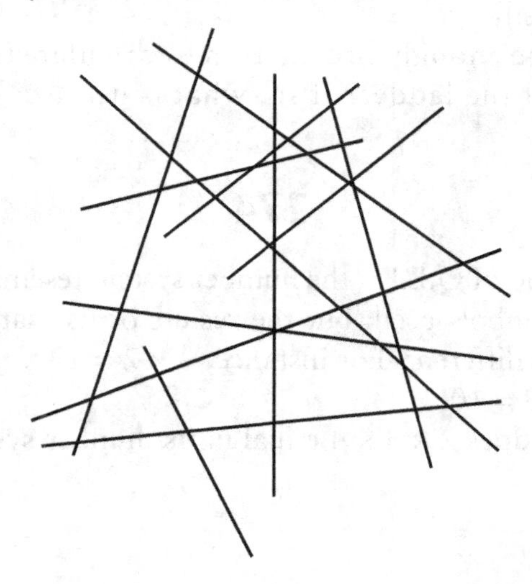

177

A boat has a speed of 20 mph in still water. It goes 40 miles downstream and comes back in 5 hours.

What is the speed of the stream?

178

A state-wide match-play golf tournament is being held, and there are 400 entrants.

How many matches, including the championship match, will be played?

179

In the stack of cubes below, imagine that it once was comprised of 64 small cubes (4 × 4 × 4). Can you tell how many cubes are missing from the original 64 cubes?

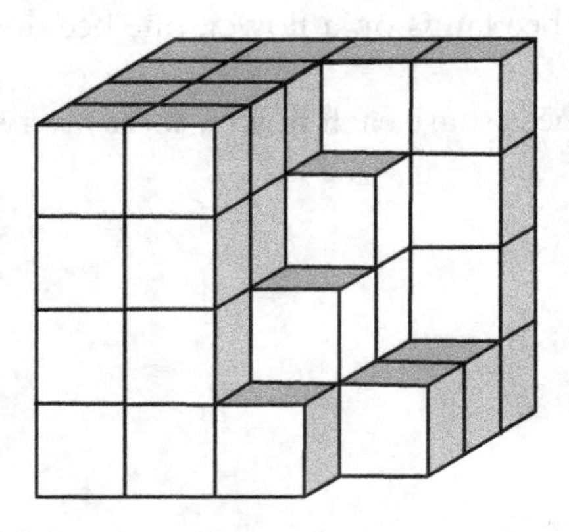

180

There is a logical sequence in the set of numbers below. Can you figure out what number should come next?

9 21 51 81 12 42

181

Without using a calculator, determine which of the following numbers equals 11!

(a) 39,916,625
(b) 39,916,800
(c) 39,815,000

182

How many flowers and bees are there if both the following statements are true?

(a) If each bee lands on a flower, one bee doesn't get a flower.
(b) If two bees share each flower, there are two flowers left out.

183

What is the minimum number of square sheets of paper — each of them unfolded, uncut, unmarked, and opaque — sufficient to create the pattern shown below, assuming that the sheets are placed flat on top of one another and that each line shown represents an edge of one of the square sheets as it has not been occluded by an overlapping square sheet?

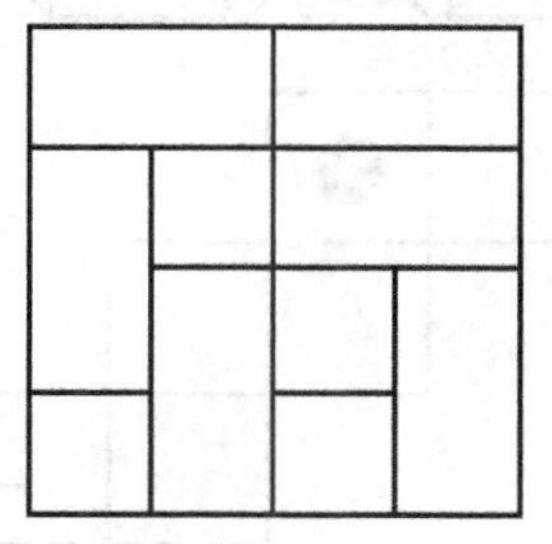

184

Player A has one more coin than Player B. Both players throw all their coins simultaneously and observe the number that come up heads. Assuming all the coins are fair, what is the probability that A obtains more heads than B?

(a) 1/4
(b) 1/2
(c) 3/4
(d) 1
(e) indeterminate

185

Below are four different views of the same cube. Fill in the three missing sides with the correct symbol and orientation.

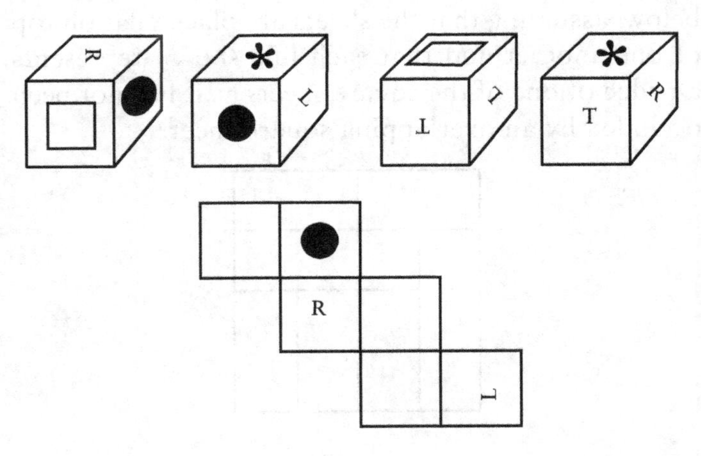

186

Let n be a positive integer. Consider the expression $\frac{n!}{10000^n}$. As n approaches infinity, the expression given approaches

(a) 0
(b) 1
(c) ∞
(d) 10,000
(e) indeterminate

187

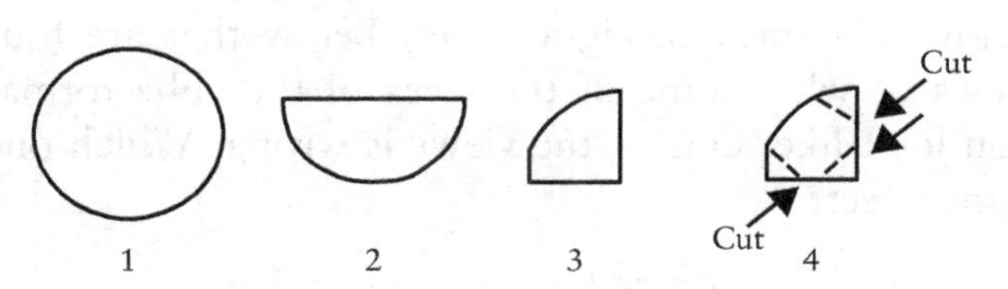

Imagine a circle cut out of a piece of paper. Fold it in half, as in Fig. 2. Fold in half again, as in Fig. 3. Then cut the tips off each of the three corners, as shown in Fig. 4. If you then were to unfold the piece of paper, it would look like:

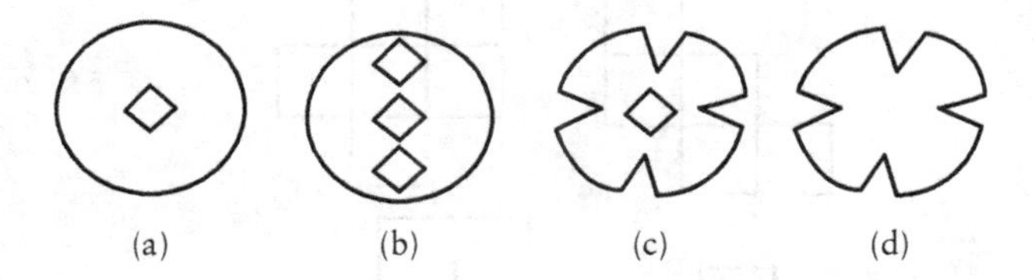

188

Below is a stack of eight cubes. Below that are four views of what some of the faces of the cube formation look like. One of the views is wrong. Which one is incorrect?

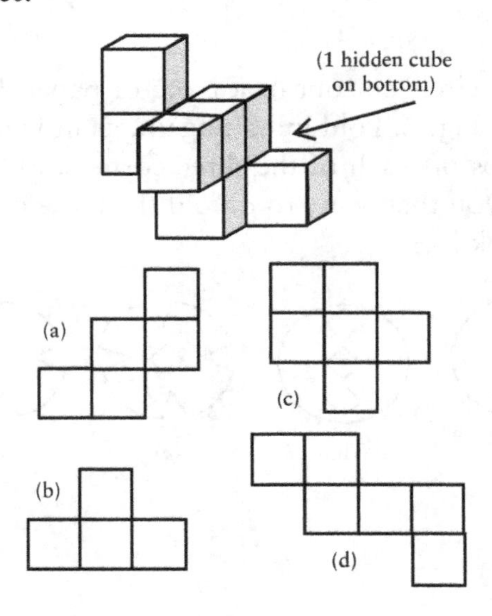

189

What is the minimum number of edges that have to be cut in a cardbord box so that it will be lie flat as shown below?

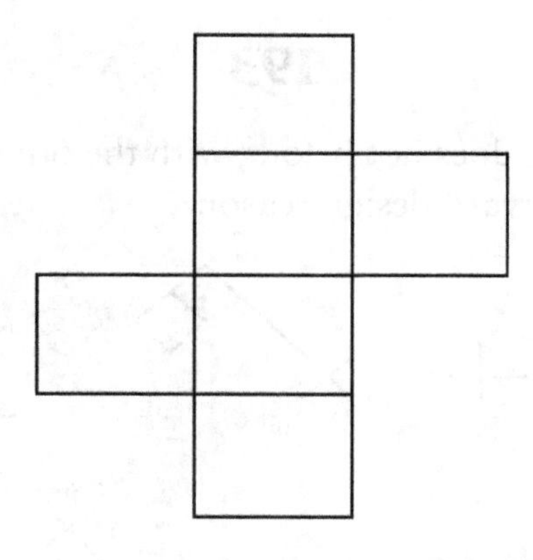

190

How many different calendars are there? By that, I mean how many calendars are different from each other, so days of the months fall on dates different from any other calendar?

191

If the area of the faces of a cube equals the volume of the cube, what is the length of one of the edges?

(Since area is expressed in square units and volume in cubic units, I'm looking for the number that is the same for each.)

192

How many integers between 4000 and 7000 have four different digits?

193

Which figure does not belong with the others based on a straightforward design reason?

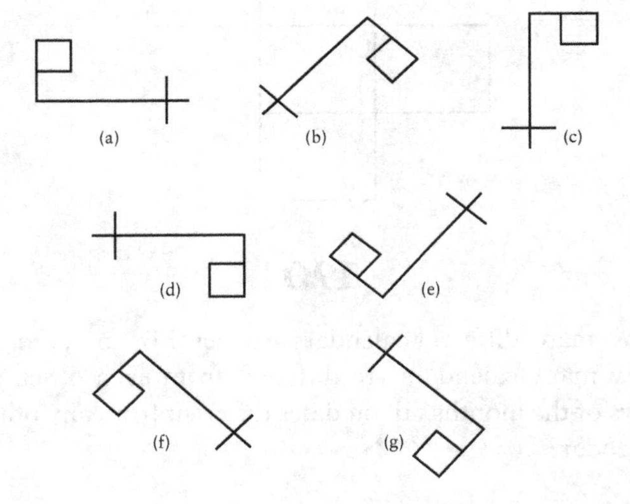

194

Which of the following five figures does not belong with the others? Symmetry is not an issue, but consider angles and how the figures were created.

There are two acceptable answers.

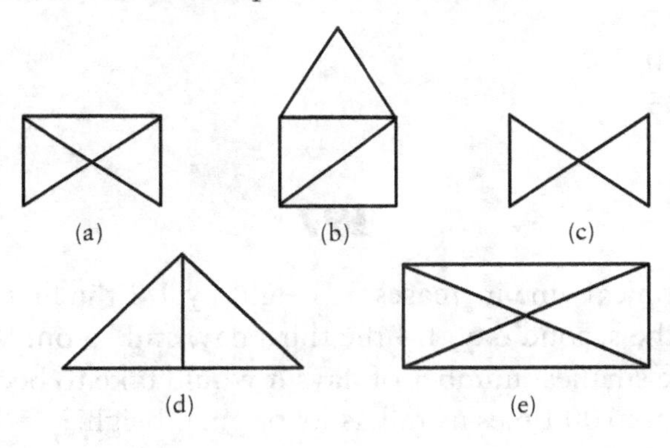

195

The total weight of a tin and the cookies it contains is 2 pounds. After 3/4 of the cookies are eaten, the tin and remaining cookies weigh .8 pounds.

What is the weight of the empty tin in pounds?

196

The quantity 12^{100} is not divisible by which of the following?

(a) 18
(b) 24
(c) 27
(d) 30
(e) 36

197

A tropical vine increases its height by 1/2 the first day, 1/3 the second day, 1/4 the third day, and so on. What is the smallest number of days it would take to become at least 100 times as tall as its original height?

198

Place five different cards in a row on a table. The third card from the left is the ace of spades.

How many different arrangements of the cards are possible if the ace of spades can never be at either end of the row?

199

(a) Which is larger? The surface area of the sphere below or the lateral area of the cylinder where the sphere sits? (The sphere and cylinder have the same radius and the same height.)

(b) Which is larger? 2^{53} or $2^{52} + 2^{52}$?

(c) Which is larger? A dozen to the power of a dozen or a dozen dozen to the power of a half dozen?

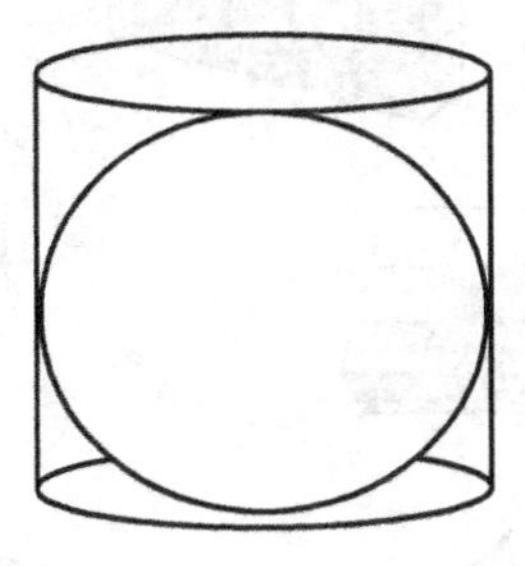

200

Three of the four figures below can be formed by cutting Fig. 1 into two pieces and reassembling them. Which figure doesn't belong?

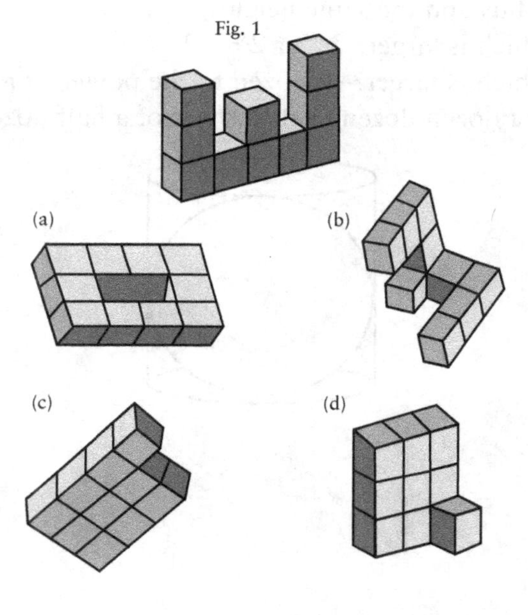

201

There is a fork in the road where one path leads to rooms filled with money and the other path leads to certain death. Guarding the fork are two brothers. One always tells the truth, whereas the other always lies. You are allowed only one question to only one of the brothers in order to make your decision as to which path to go down. What would your question be?

202

A container holds 1500 ml of a salt solution. How many ml of water must be added to decrease the saltiness by 30 percent?

203

Ed rowed one mile downstream in three minutes. Rowing with the same force upstream, it took him four minutes to cover one mile. How much time would it take Ed to row one mile in perfectly calm water?

204

If Cherie's cousin's cousin's son is Cherie's nephew, then this nephew's father could be Cherie's________.
(a) father
(b) brother
(c) grandfather
(d) there is no possibility of them being related

205

A cab driver picks up the same person at the same time every morning and takes her to work. One morning she decided to leave home one hour earlier, walking to work along the same route as the cab takes. She notifies the cab driver she will be walking but tells him she may call him if she gets tired. After walking some distance, she calls the cab to pick her up and continue to the office. She reaches the office 20 minutes earlier than usual. How long had she been walking?

206

Bill and Bob work for a grocery store and each is carrying a partially full crate of pop (a full crate would hold 24 cans).

Bill says to Bob, "If you give me a certain number of cans from your crate, I'll have 3 times as many cans as you. But, if I gave you the same certain number, you would have exactly 1/2 of what I have." How many cans of pop does each have and what is the "certain number" they would have to give each other?

207

While driving down an unfamiliar road one night, I noticed a strange sign that read "Entropy – 150 *ents* or, using the Tesseract System, the distance is 110 *tesses*." Further down the road I saw "Entropy – 10 *ents* or 26 *tesses*." Clearly I had broken through to a different dimension into what physicist call "Puzzleland Chaos."

My mind began to wander... was there a point down the road where the distance marker measured in *ents* was equal to the distance marker measured in *tesses*? I am referring to the marker only, not the true total distance, which would be different for both.

208

I had some baseball cards with me. I gave half of them with one more to my daughter. Then, I gave half of the remaining cards with one more to my son. Now I was left with just one card. How many baseball cards did I have in the beginning?

209

Alex can beat Beth by 8 yards in a 200-yard race. Beth can beat Christy by 6 yards in a 200-yard race. If Alex were to race Christy in a 50-yard race, how many yards would have been completed by Christy at the instant Alex crossed the finish line?

210

What is the smallest number, when spelled out, that uses the five vowels: a, e, i, o, u, in alphabetical order?

211

Which is larger $-2^{-3^{-4}}$ or $-4^{-3^{-2}}$?

212

Bill's wife's mother-in-law is Bill's wife's sister's sister-in-law and Bill's brother-in-law is Bill's sister's stepfather. How did this happen?

213

A pilot has a student copilot and they are on a straight line flying lesson. They left Albion and are flying due East to Clarion. A town called Blanton is somewhere between Albion and Clarion.

After some time passes, the student asks the pilot, "How far are we from Albion? It seems like we've been flying for a long time." The pilot responds, "One third as far as from here to Blanton."

A while after passing Blanton, they are 600 miles from where the student asked the question. He now asks a second question: "How far are we from Clarion?" The pilot smiles and says, "One third as far as from here to Blanton."

How far is Albion from Clarion?

214

What is the expected number of coin flips for getting two consecutive heads?

215

How many randomly assembled people are needed to have a better than 50 percent probability that at least one of them was born in a leap year?

216

Using four 2s and one math symbol of your choosing, what is the largest number that can be created?

217

Determine the relationships between the figures and the words to find solutions to the two unknowns.

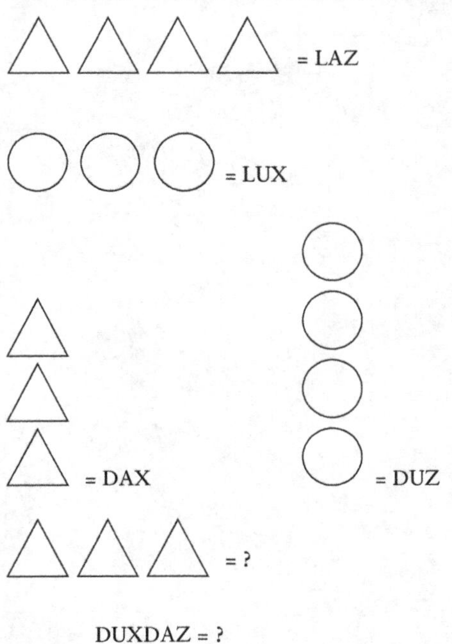

DUXDAZ = ?

SOLUTIONS

1

Pulley A is moving in the opposite direction.

2

There are 26 cubes.

3

The bucket weighs 2 lbs.
Here's one way to think about this:
If x is the weight of water and y is the weight of the bucket, then

$$1/2x = 25 - 13.5$$
$$= 11.5.$$

So $x = 23$ and $y = 2$.

4

(b) 8. Here are the possible combinations:

11 19 23→53 – 1
13 19 23→55 – 2
17 19 23→59 – 3
11 19 29→59
13 19 29→61 – 4
17 19 29→65 – 5
13 19 31→63 – 6
17 19 31→67 – 7.

5

I would have to deal you 45 cards to guarantee you have a straight with 100% accuracy.

6

5:00 p.m.

If it were three hours later, it would be 8:00 p.m., which is four hours until midnight.

One hour earlier would be 4:00 p.m., which is 8 hours before midnight. So the four hours from 8:00 p.m. is one half the time from 4 p.m. until midnight.

7

Here's one way to look at this puzzle:

a) $\dfrac{1/3}{7} = \dfrac{5/16}{x}$

b) $1/3x = 35/16$

c) $x = 105/16$

8

Because the square has side 35, 15 + C + 12 = 35, so C = 8. A is 7, and B = 3.

9

Opposite the blackened square ■
is the blackened diamond ◆ and
opposite the transparent circle ○
is the blackened circle ●.

10

The answer is (b) 9. Because $2^{2^n} = (2^2)^{4^4} = (2^2)^{256} = 2^{512}$, $2^n = 512$ and $n = 9$.

11

1/3

The possibilities are

(H) (H)

~~(T)~~ ~~(T)~~

(H) (T) (dime is heads, penny is tails)

(T) (H) (dime is tails, penny is heads).

Since TT can be eliminated, you can see there is only a one in three chance both coins are heads.

12

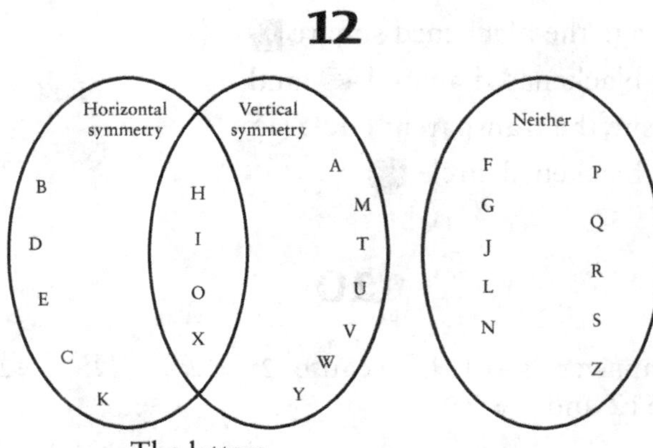

The letters
B C D E K
have reflection
across the horizontal
symmetry plane.

13

There would be only two 5s. 10! = 3,628,800. There is no multiple of 5 higher than 25 that will divide evenly into 10!

For 15! 125 will divide into it evenly, so there would be three 5s.

14

The answer is (e). The key is the number of supporting wires. Each has three different wires or ropes supporting the weight.

15

For no sliding,

$$f(a + z) = b.$$

So

$$.2(20 + z) = 10,$$
$$4 + .2z = 10,$$
$$.2z = 6,$$

and

$$z = 30.$$

16

15 women
6 men

Let x be the number of men. Then there are $x + 9$ women and we are given that

$$\frac{x+9}{x} = \frac{5}{2}.$$

Solving,

$$2(x + 9) = 5x,$$
$$2x + 18 = 5x,$$
$$3x = 18,$$

and

$$x = 6.$$

So there were 6 men and 15 women.

17

You would need two each of 1, 3, 9, and 27 ounce weights. You can balance any whole number of ounces between 1 and 40 with these eight weights.

18

The answer is 10.

These are the fractions 1/2, 1/3, 1/4, 1/5, 1/6, 1/7, 1/8, 1/9, and 1/10 expressed in decimal form (without the decimal), and rounded off.

1/2 = .5	1/6 = .1666...	
1/3 = .3333...	1/7 = .142857...	1/10 = .10
1/4 = .25	1/8 = .125	
1/5 = .20	1/9 = .111...	

19

If $g = 1 - \sqrt[t]{v/f}$ then

$$v = f \times (1 - g)^t,$$
$$(1 - g)^t = v/f,$$
$$1 - g = \sqrt[t]{v/f},$$
$$-g = -1 + \sqrt[t]{v/f},$$
and $g = 1 - \sqrt[t]{v/f}$.

20

A plane is determined by three non-collinear points. So, if we are given five points and no three points on the same line, then there are five choose three or ten planes that can be formed. Assuming every pair of planes intersect then there would be ten choose two maximum intersecting lines, or forty-five intersections, maximum.

21

A = 9, B = 3, C = 7, D = 2.
12 + C + 9 = 28, so C = 7.
Then 8 + 7 + D + 11 = 28, so D = 2.
Because 12 + B + 8 + 5 = 28, B = 3.
From 12 + A + 7 = 28, A = 9.

22

The answer is 14 days.

There were 15 half days (6 + 9) where no chocolate was eaten and 13 half days when it was, which means 15 – 13 = 2 half days or one full day when no chocolate was eaten. 13 + 1 = 14.

23

The probability that of seven people chosen at random, one or more was born on a Friday is .6601 and the probability that exactly one person was born on a Friday is .397. In the first case, six out of seven people can be born on any other day of the week. So, for seven people that would be $\left(\frac{6}{7}\right)^7$, which is .3399. The certainty that someone is born on a Friday is 1. Therefore, $1 - .3399 = .6601$.

In the second case, the probability is $\left(\frac{6}{7}\right)^6$ or .3966. If only the first person were born on Friday, the probability is $\frac{6^6}{7^7}$. But the probability is the same for each of the seven. So, $7x\left(\frac{6^6}{7^7}\right)$ or $\frac{6^6}{7^6}$, which is .3966 (.397).

24

The answer is 20.

25

The answer is –60.

The sequence goes up 4, down 8, up 8, down 16, up 16, down 32, up 32, down 64, and $4 - 64 = -60$.

26

The answer is (d).

27

Here is one solution but there are probably other, shorter solutions:

(a) From the 12-cup container, fill the 7-cup container giving you 5, 7, and 0 cups
(b) Fill the 6-cup container from the 7-cup container giving you 5, 1, and 6 cups
(c) Now take the six cups and pour back into the 12-cup container giving you 11, 1, and 0 cups
(d) Move the one cup in the 7-cup container to the 6-cup container giving you 11, 0, and 1 cups
(e) Fill the 7-cup container with the 11 cups in the biggest container giving you 4, 7, and 1 cups
(f) Fill the 6-cup container from the 7-cup container giving you 4, 2, and 6 cups
(g) Put the six cups from the smallest container into the largest container giving you 10, 2, and 0 cups
(h) Move the two cups into the smallest container giving you 10, 0, and 2 cups
(i) Fill the 7-cup container from the largest container, which now holds 10 cups giving you 3, 7, and 2 cups
(j) Fill the 6-cup container from the 7-cup container giving you 3, 3, and 6 cups
(k) Finally, fill the 12-cup container with the six cups from the small container and you have nine cups – 9, 3, and 0 cups.

28

The answer is (d). It is the only figure that does not have a 90° angle.

29

196.5 feet per second (about 134 mph). Let x be the speed, in feet per second. Moving to 90 ft. and having the ball reach home plate in 0.458 seconds, we have $90/x = .458$, so $.458x = 90$, $x = 90/.458 = 196.5$.

30

17 goes in the center circle, and 6 goes in the corner. In each rectangle, multiply the top two numbers. That results in a 2-digit number. Put that number into the two bottom corners, one digit per corner. The center number in each rectangle is the sum of the four corners.

31

The answer is A = 48 lbs.: B = 72 lbs.: C = 116 lbs.

Take a look at the 4 ft. segment supporting 30 lbs. 4 ft. × 30 lbs. = 120 ft. lbs. If there are 120 ft. lbs. on the right side of this portion of the mobile then there must be 120 ft. lbs. on the left side — the side supported by 2 ft. We know 60 lbs. has to be distributed in a 2 to 3 ratio on the left side. 2 ft. × 60 lbs. = 120 ft. lbs. So we know it will balance with the right side. And since it must be in a 2 to 3 ratio, 48 lbs. will be the weight in A and 72 lbs. in B.

Now, we know this portion of the mobile balances because 3 ft. × 48 lbs. = 144 ft. lbs. = 2 ft. × 72 lbs. Now add all the weight on the left side of the mobile — the weight supported by the 2 ft. segment at the very top. 200 + 30 + 60 = 290 lbs. 290 lbs. × 2 ft. = 580 ft. lbs. on the left side. We need the same on the right side supported by 5 ft. at the top. 5 times some number equals 580.

$5x = 580 \qquad x = 116$

So, 116 lbs. is the weight of C.

32

20 students were needed. 48 days times 5 students (per shift) means they needed 240 total shifts. But each student could only work 12 shifts, so 20 students were needed.

33

There are 46 houses total on both sides of the street. There are several ways to solve the problem. One way is to realize the total number of houses will always be one less than the sum of two opposing houses. Another is to see there are 22 houses between houses 12 and 35. That means 11 houses have to be beyond house 12 on that side of the street. So that side has 23 houses and there must be double that number, or 46, total.

34

Areas 1 and 2 are the same. Let each of the four small circles have an area of x. Let a and b stand for the areas of 1 and 2, respectively. The area of the smaller circles is $4x - 4a$. It is also equal to the area of the large circle, minus $4b$.

So,
$$4x - 4a = 4x - 4b$$
$$-4a = -4b$$
and $a = b$.

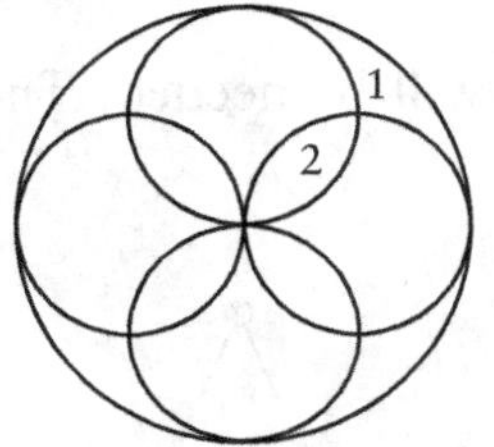

35

The answer is 2/3.

Here is one way to consider the problem: Label the balls A, B, C, and D. Let A, B, and C be the black balls. You picked a black ball. It can either be A, B, or C with equal probability (1/3). Since in two of those three cases (when you pick A or B), you would end up with another black ball in the next draw, your probability would be 2/3.

36

There are 57.87 days in 5,000,000 seconds.

5,000,000/60 = 83,333.33 minutes,
83,333.33/60 = 1,388.88 hours,
and 1,388.88/24 = 57.87 days.

37

Only ten bushes will be needed. They can be planted as follows:

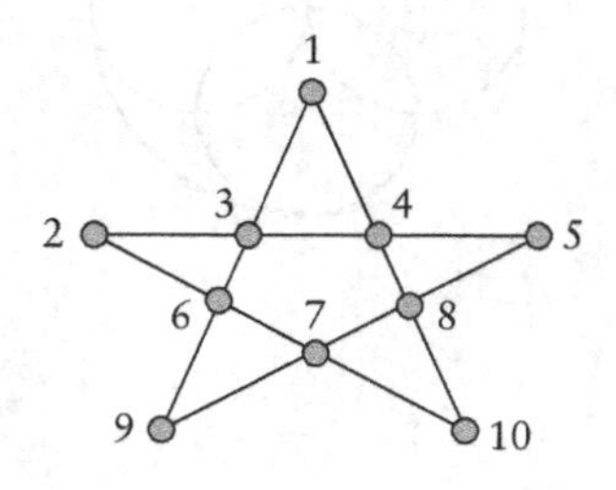

38

The answer is (c) 3.

Here's one way to solve the problem. Start by working from the inside.

$27^{4/3}$ is the cube root of 27, taken to the 4th power. That is, $3^4 = 81$, so
$81/3^2 = 81/9 = 9$,
and $\sqrt{9} = 3$.

39

The answer is 500 lbs.

If 95% of 1,000 pounds was originally water, then 50 pounds was not. (5% of 1,000 is 50.) At the end of the trip, there is still 50 pounds of material that is not water, and it now accounts for 10 percent of the load, so the load is 500 pounds.

40

Six typists.

If two typists can type two pages in two minutes, then each one can type one page in two minutes, so he or she can type half a page in a minute.

A typist who can type half a page in a minute can type (0.5×6) pages in 6 minutes, or 3 pages a minute. But we have 18 pages to type, so we need (18/3) typists or six typists.

41

11^{-11}.

Did you find a smaller number?

42

1) (c) $2^{3/2}$; 2) (c) $3w - q - p$; 3) (c) 125/512.

43

The result would look like (e).

44

The answer is 6 2/3 hours.

Let r be the rate in gallons per hour of the slower pump and let V be the volume of the pool.

We know

$$(r + 1.5r) \times 4 = V,$$
$$\text{so } 10r = V.$$

Thus

$$(1.5r) \times 20/3 = V,$$

So the faster pump would take 20/3 hours to fill the pool.

45

You would need 14 gallons of paint.

There are 42 external sides (the same number as the number of faces on 7 cubes). Since two gallons are needed to paint one cube, you would need 2 × 7 or 14 gallons of paint to cover the figure.

46

The answer is 1096.

Notice the differences between the consecutive numbers. Each successive number is three times the previous number.

4		7		16		43		124		367		1096
	V		V		V		V		V		V	
	3		9		27		81		243		729	

47

Removing ten matchsticks will leave no squares of any size.

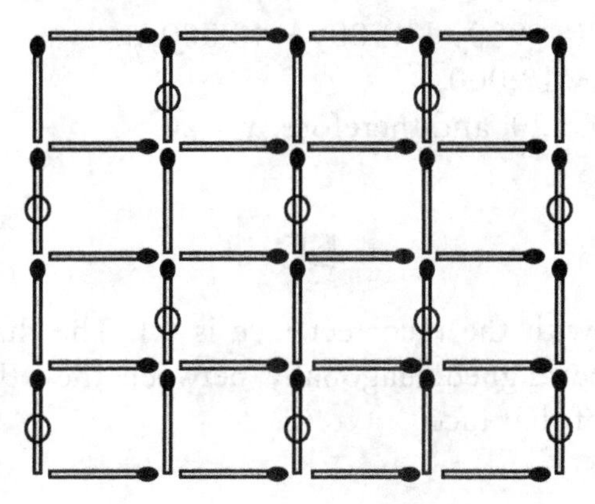

48

The answer is 1,320 miles.

Let the distance traveled by the first train be $x + 120$ miles, and the distance covered by the second train be x miles. Then

$$\frac{x}{50} = \frac{x+120}{60},$$

$x = 600$,

and so the distance between A and B is

$x + x + 120 = 1{,}320$ miles.

49

There will be 20 pages of 900 words and 20 pages of 1,500 words. Here's one way to view this. We know

$900x + 1500y = 48{,}000$ and $x + y = 40$

Then $x = 40 - y$ and substituting,

$900\,(40 - y) + 1500y = 48{,}000.$

That is,

$36{,}000 - 900y + 1500y = 48{,}000$, so

$600y = 12{,}000$,

and $y = 20$, and therefore, $x = 20$

50

The die with the incorrect face is (c). The three dots should be aligned diagonally between the other two corners of that face.

51

The unfolded cube is

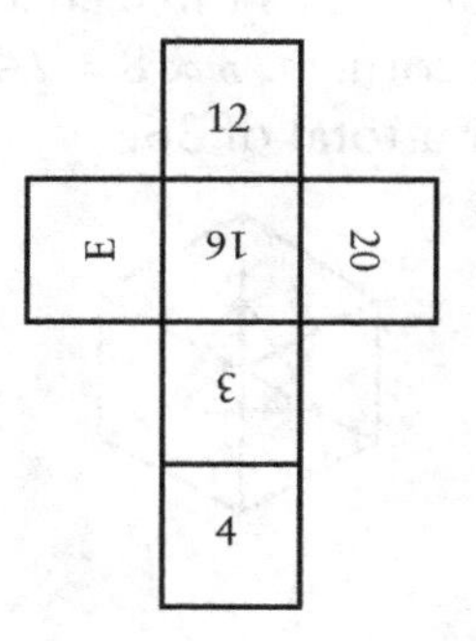

52

The answer is (b) 4 pounds.

If 4/5 of a pound of cheese is equivalent to 1/5 of a block of cheese, then 5(4/5) = 4 pounds will be equivalent to 5(1/5) = 1 block.

53

The answer is (c) 8.

54

The answer is 36. One way to think about this is to see that each sliced corner has three edges.

There are eight corners. $3 \times 8 = 24 + 12$ (number of original edges) for a total of 36.

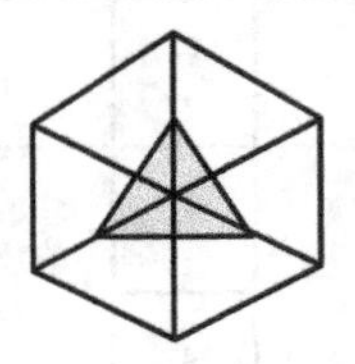

55

(a) The first fraction is 0.31623 and the second fraction is 3.1623. The second is larger.
(b) The same
(c) The same

56

The answer is (d) $2^{96} + 1$.
Let $2^{32} = x$. Then $2^{32} + 1 = x + 1$.
Let $x + 1$ be divisible by the number n.
Then, $2^{96} + 1 = [(2^{32})^3 + 1] = x^3 + 1 = (x + 1)(x^2 - x + 1)$, which is exactly divisible by n, since $(x + 1)$ is divisible by n.

57

The segment in the lower right of the last number is used more than any other.

58

There are 96 cubes.

59

The answer is (d) 1014.
If there are 15 cubes to each side, then a grid of 13×13 will have one face painted on each side of the cube, and $13 \times 13 \times 6 = 1014$.

60

The answer is 5 because there are 105 Leeps in a Zank and 21 Grats in a Zank.

61

G is in the eighth position.

The complete order is

D E C H A B F G I

62

The answer is 1/8.

$4^x \cdot 4^{-2x} \cdot 4^{3x} = 4^{1/4}$ so

$x - 2x + 3x = 1/4$,

$2x = 1/4$,

and $x = 1/8$.

63

Thirty students are majoring in both. If 72 percent of the 100 students are business majors, then 28 percent are not. If 58 percent are math majors, then 42 percent are not. So 28 + 42, or 70 percent, are not majoring in both.

64

The unfolded cube is

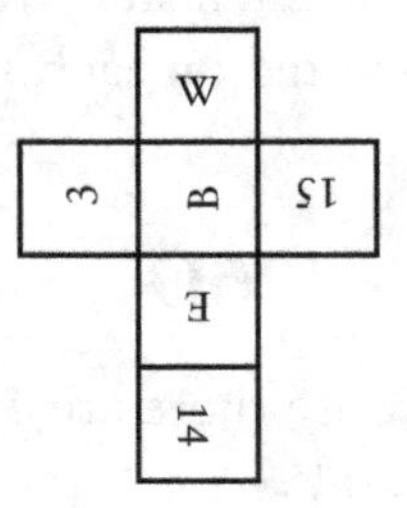

65

5-4-4 is not allowed but 6-4-3 is, so the smallest amount of coins that can be put in the largest pile is six.

66

PEACE occurs in the first and second message and so has to be FLANG.

IN occurs in the first and third, so it has to be BRONO. So, HASSI translates to WE COME.

67

Two ways are $512 \times 125 = 64{,}000$ and $2^9 \times 5^3 = 64{,}000$. Did you find another?

68

The answer is 56 boys.

If $2x$ boys attend the school, then $7x$ girls do.

We are given that

$$\frac{2x+8}{7x} = \frac{1}{3}$$

so $7x = 3(2x + 8) = 6x + 24$ and $x = 24$.

Thus $2x + 8 = 56$.

69

The answer is 6 dots.

The faces and their opposites are 3 and 6, 2 and 4, and 1 and 5.

70

Each of the hash marks represents 1 foot.
On the top left side, $60 \cdot 2 + 20 \cdot 3 = 180$.
On the top right side, $50 \cdot 2 + 80 \cdot 1 = 180$.
On the second level, the 40 weight is balanced by the balanced 10 and 30 weights: $(10 \cdot 3 = 30)$.

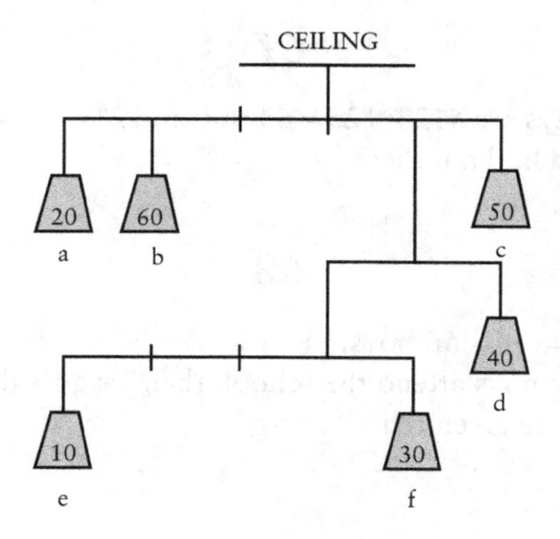

71

Molly should be driving a car that has a jet engine and wings! She will need to cover the final 21 miles at 190.91 mph. If Molly needs to cover the whole trip averaging 19 mph, she will need to cover the 42 miles in 42/19 or 2.21 hours. However, she has already covered the first half of the trip (21 miles) in 2.1 hours (21/10). She needs to cover the last 21 miles in 2.21 – 2.1 or .11 hours and 21/.11 = 190.91 mph.

72

It is an odd integer. The powers of 2 form a pattern: it can be seen that 2^{58} would end in 4. Adding one to that number will result in a number ending in 5. When that number is divided by 5, the result is an odd integer.

$2^1 = 2$, $2^2 = 4$, $2^3 = 8$, $2^4 = 16$, $2^5 = 32$, $2^6 = 64$, $2^7 = 128$, and $2^8 = 256$

73

The answer is Tuesday. Work from the end backwards. "The day before tomorrow" is today, Monday. "Two days before" that is Saturday. "Three days after" that is Tuesday.

74

The numbers are 20 and 30.
Let the numbers be x and y. Then
we are given

$$x + y = 50$$

$$\frac{1}{x} + \frac{1}{y} = \frac{1}{12}.$$

Substitute $y = 50 - x$ in the second equation:

$$\frac{1}{x} + \frac{1}{50-x} = \frac{1}{12},$$

$$12(50 - x + x) = x(50 - x),$$

$$600 = 50x - x^2,$$

$$x^2 - 50x + 600 = 0,$$

$$(x - 30)(x - 20) = 0,$$

$$x = 20, y = 30,$$

so the numbers are 20 and 30.

75

The answer is 6 dots.

Opposite faces on normal dice always total seven.

76

Only one is an orthodontist to go along with the 99 oral surgeons.

77

There are 600,001 ones between 1 and 1,000,000 (inclusive). Here is the pattern of occurrences between

1 and 10 (inclusive) — 2
1 and 100 (inclusive) — 21
1 and 1000 (inclusive) — 301
1 and 10,000 (inclusive) — 4,001
1 and 100,000 (inclusive) — 50,001
1 and 1,000,000 (inclusive) — 600,001

78

The height must be reduced by 59.1%.

Increasing the diameter by 30%, is the same as multiplying it by 13/10, so the area of the base is increased by $(13/10)^2 = 169/100$.

To keep the volume the same, the height must be

$$\frac{1}{169/100} = \frac{100}{169} = 59.1\%.$$

79

\$89.

Multiply the two values and subtract their sum: $10 \times 11 = 110$, $10 + 11 = 21$, $110 - 21 = 89$. In general, the largest sum that cannot be gotten as a combination of x and y is $xy - x - y$.

80

The answer is (c) 11/36.

Count the possibilities. Of the 36 ways a die can be tossed twice, 11 result in at least one six.

81

The answer is 7.5 hours.
Her new time is $4/3t$. So $4/3t = t + 5/2$,
$1/3t = 5/2$,
$t = 15/2$.

82

The horizontal symbols both contain the letter B, therefore B stands for figures arranged horizontally. S appears only in vertical arrangements. The letter O always appears with the * figure, so O stands for *. I stands for the ○ symbol. P stands for the number 3, and Z stands for 2.

So * * = BOZ

SIZSOP = ○
○
*
*
*

83

The answer is (b) 1 out of 65.

The probability of rolling 1, 2, 3, 4, 5, and 6 with six dice in a single roll is $1 \times (5/6) \times (4/6) \times (3/6) \times (2/6) \times (1/6) = 0.015432$. This is the same as $6!/6^6$, which comes out to be 1 out of 64.8.

84

Although we all know it is impossible to lay a fraction of an egg, mathematically, each hen will lay 2/3 of an egg in one day.

If a hen and a half lays an egg and a half in a day and a half, then one hen lays one egg in 1.5 days. 1.5 days is 3/2 days.

1 hen × 3/2 days × rate per hen per day = 1 egg so the rate has to be 2/3.

85

The answer is 16.

36 + 8 must equal 20 plus the number in the middle because both sets share the number that goes in the upper left corner. That makes the middle number 24. The second row is 8, 24, and 40, which total 72. So the third row will be 36, 16, and 20 (which total 72).

86

The numbers are 6, 10, and 14.

Since we have is an arithmetic sequence, the difference between successive numbers is constant. Thus
$(5x - 10) - (2x - 2) = (3x + 2) - (5x - 10)$,
$5x - 10 - 2x + 2 = 3x + 2 - 5x + 10$,
$3x - 8 = -2x + 12$,
$5x = 20$ and $x = 4$, so
$2x - 2 = 6$,
$5x - 10 = 10$,
and $3x + 2 = 14$.

87

There are eight different ways.

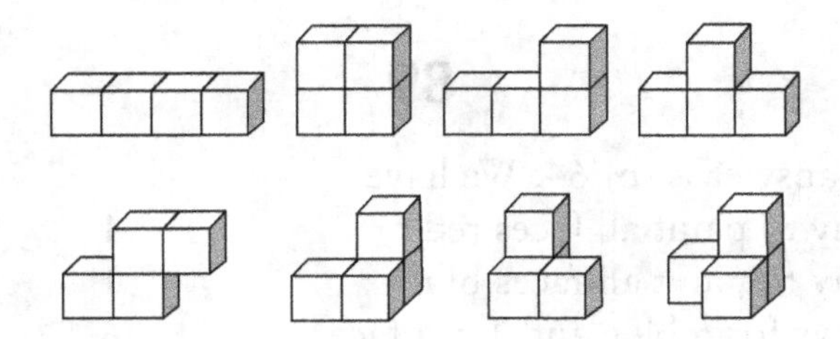

88

The answer is 48 seniors.
Let x be the number of juniors. We were given

$$\frac{3}{8} = \frac{x}{x+40}$$

so

$$3(x + 40) = 8x$$
$$120 = 5x,$$

and

$$x = 24.$$

At the start there were 24 juniors and 64 seniors. One-half of the juniors left the club to join the activity, which means 12 left and 12 remained. That means 16 seniors stayed in the club: 12/16 = 3/4. Thus 64 – 16 = 48 seniors left the club to join the activity.

89

The answer is (b) 64. We have

1 way to paint all faces red	1
1 way to paint all faces blue	1
6 ways for 5 blue and 1 red face	6
15 ways for 4 blue and 2 red faces	15
20 ways for 3 blue and 3 red faces	20
15 ways for 2 blue and 4 red faces	15
6 ways for 1 blue and 5 red faces	6
	64

90

The answer is 4 hours and 45 minutes.

If Linda made two round trips the first way, it would take her 6 hours. That is, she would cover the distance twice by walking and twice by riding. So 6 hours minus 1 hour and 15 minutes will be the total time walking both ways — 4 hours, 45 minutes.

91

(c) is the only one that could be true because it shows the other statements to be false.

92

The answer is 94.37 percent.
Each number is now .75 of its original value. So $(.75w)\,(.75x)^2\,(.75y)^4\,(.75z)^3 = (.75)^{10}\,wx^2\,y^4\,z^3 = .0563$ $wx^2\,y^4\,z^3$, and $1 - .0563 = .9437$.

93

There are 40 cubes.

94

There are 25 squares.

95

The answer is 3 3/7 miles. Here's one way to solve the problem where x is the number of miles. Then

$x/3 + x/4 = 1,$

$4x + 3x = 12,$

$7x = 12$, and $x = 12/7$.

The total distance is $2x$, so $2(12/7) = 3\ 3/7$ miles.

96

The answer is 626 pages. There are the digits 1–9 which are pages 1–9. Then the pages 10–99 account for 180 digits. So for the first 99 pages, we have 189 digits. Pages 100–999 have 3×900 or 2700 digits. But we only have 1770 digits total, and we have already accounted for 189 of them (99 pages).

If x is the number of three-digit pages, then

$189 + 3x = 1770$

$3x = 1581,$

and

$x = 527.$

Thus the total number of pages is $527 + 99 = 626$ pages.

97

4500 oz. of pure water.

Here's one way to find this:

The container has $1500(.4) = 600$ ounces of salt. For this to be in a 10% solution, there has to be 6000 oz of water. So $6000 - 1500 = 4500$ ounces of water needs to be added.

98

The answer is .0667%.

15 percent of 1/15 is .01 and
.01/15 = .0667%.

99

The number of triangles is 35. The triangles are

ABC	ADI	BDE	DEF
ABD	AEF	BDF	DEH
ABE	AEG	BEJ	DEI
ABF	AEH	BGI	DHI
ABG	AFG	CDE	EFH
ABI	BCD	CDH	
ACD	BCE	CDI	
ACE	BCG	CDJ	
ACH	BCI	CEG	
ADE	BCJ	CIJ	

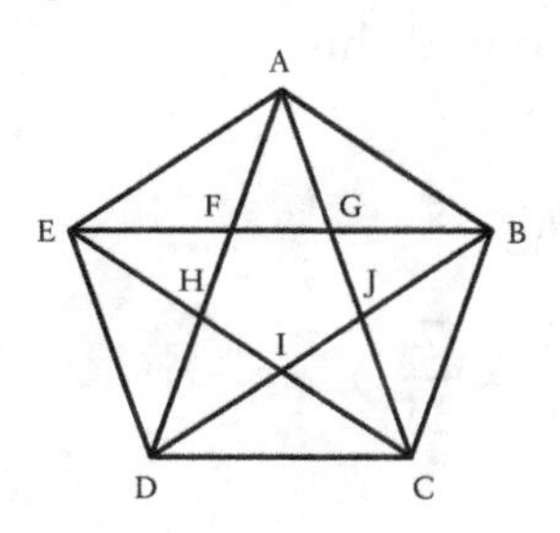

100

The answer is (c) 4.

$$\left(\frac{1}{5}\right)^{10} = \frac{1}{5^{10}} = \frac{2^{10}}{2^{10} \cdot 5^{10}} = \frac{1024}{10^{10}} = .0000001024.$$

101

The answer is 8 p.m.

From 1 p.m. to 1 a.m. is 12 hours. So 4 hours after 1 p.m. is 1/2 of the 8 hours before 1 a.m. That makes the time, at that point, 5 p.m. Since this occurred 3 hours ago, it is now 8 p.m.

102

The answer is (a) 7/25.
Here's one way to find this.

$\sqrt{7}$ percent is $\frac{\sqrt{7}}{100}$, so

$$\frac{\sqrt{7}}{100} \times 4\sqrt{7} = \frac{28}{100} = \frac{7}{25}.$$

103

There are 20 ways. Five numbers may be chosen $5 \times 4 \times 3 \times 2 \times 1$ ways, but three of them are the same. We don't count the duplicates of repeating numbers. Since there are three repeating digits, there are $3 \times 2 \times 1$ or 6 ways of ordering them, so we have

$$\frac{5 \times 4 \times 3 \times 2 \times 1}{3 \times 2 \times 1} = \frac{120}{6} = 20 \text{ ways.}$$

They are

68999	89699	96998	99698
69899	89969	98699	99869
69989	89996	98969	99896
69998	96899	98996	99968
86999	96989	99689	99986

104

The least number of cubes is seven.

105

Martha and Mike played the 11th set.

The total number of sets played was

$$\frac{(15 + 14 + 9)}{2} = \frac{38}{2} = 19.$$

Minnie played 9 sets and sat out 10 sets, which means she lost the even-numbered sets, leaving Martha and Mike as the players of the 11th set.

106

Brian is the first baseman.
Steve is the second baseman.
Rich is the shortstop.
Kendall is the third baseman.
Rich and Steve are brothers.
Kendall is Steve's son and Rich's nephew.

107

The answer is 21 (one 5×5 square, seven 2×2 squares, three 3×3 squares, and ten 1×1 squares).

108

Let x be the number of cashews and y the number of peanuts.
Then $8x + 3y = 6(x + y)$, but we know $x + y = 20$, so $8x + 3y = 6(20)$ or 120.

To solve $8x + 3y = 120$
$x + y = 20$

Multiply the second equation by –8, so

$$8x + 3y = 120$$
$$-8x - 8y = -160$$

Adding,
$-5y = -40$ so
$y = 8$, and therefore $x = 12$.

The answer is 8 pounds of peanuts and 12 pounds of cashews.

109

The answer is 2 to 3. Let the bicycle's current age be $3x$, making the tires' age x when the bicycle was as old as the tires are now. To make them the same age, we must add to the tires' age some number, y, and subtract from the bicycle's age the same number, y. Thus

$$3x - y = x + y \text{ so}$$
$$2x = 2y \text{ and}$$
$$x = y.$$

Knowing that $x = y$, we can substitute y for x in the bike's current age: $3x = 3y$

The tires' current age is then $2y$, and the ratio of the tires' current age to the bicycle's current age is $2y/3y$, a ratio of 2 to 3.

110

They are not the same. Figure (d) is different.

111

(a) If any two of the beginning colors are the same number, such as 15 red, 18 blue, 15 green, then the reds and greens meet to form 30 new blues, and the total becomes 48 blues (30 + 18) or this situation:

(b) Say you have 5 blue, 2 green, and 1 red (8 total). Match one blue with one red, resulting in two new greens and leaving no reds. We now have 4 greens and 4 leftover blues. Now match the 4 blues with the 4 greens and you will end up with 8 reds.

Can you come up with an alternate solution involving fewer than 20 chameleons?

112

The largest number is $5^{6^{7^{8}}}$ which is approximately 10 to the 4.5 millionth power. It is the largest number by a long way. Second is $5^{6^{8^{7}}}$. Third is $5^{7^{6^{8}}}$.

113

There are 18 different sides or faces.

114

The answer is 60 mph.

It takes Molly 21/12 or 1.75 hours to go the first 21 miles. For her to average 20 miles an hour for 42 miles, she needs to cover those 42 miles in 42/20 or 2.1 hours. That means she has to cover the last 21 miles in $2.1 - 1.75$ or .35 hours. Because $21/.35 = 60$, she will have to average 60 mph on the second half to average 20 mph for the entire trip.

115

The answer is 15 hours. If t denotes time in hours, we want to solve $7{,}500 - 150t = 4{,}500 + 50t$.

That is $3{,}000 = 200t$, so $t = 15$.

116

The answer is 10 percent.

We're given that

$$.001x = .1y$$

so $$x = 100y$$

or $$\frac{x}{10} = 10y.$$

So $10y$ is 10% of x.

117

The answer is 100 steps.

Let Andrea walk down 1 step per second and suppose that the escalator moves n steps per second. It takes Andrea 50 steps to reach the bottom in the same amount of time that the escalator steps would have covered $50n$ steps. The total steps are then found to be $50 + 50n$. Marty takes 90 steps to reach the bottom in the same time it takes Andrea to cover 10 steps — or 10 seconds. So in 10 seconds the escalator would have covered $10n$ steps. Marty's total is then $90 + 10n$. So $50 + 50n = 90 + 10n$. That gives $n = 1$ and $50 + 50(1) = 100$ steps.

118

The answer is (a) 5.

$$.2^{-2} = \frac{1}{.2^2} = \frac{1}{.04} = 25 \text{ and } \sqrt{25} = 5.$$

119

Central Michigan would beat Michigan by 15 points. If Minnesota were to play Oklahoma, they would win by 29 points. Nebraska beat Oklahoma by 4 and since Minnesota would have beaten Oklahoma by 29, they would beat Nebraska by 25. Michigan beat Nebraska by 14, so Minnesota would beat Michigan by 11. Central Michigan beat Minnesota by 4 points, so Central Michigan would beat Michigan by 15 points.

120

There were 75 physicians total at the luncheon.

If the number of physicians is n, then we are given:

The number of neurosurgeons is $n - 40$.

The number of pediatricians is $n - 50$.

The number of cardiologists is $n - 60$.

Their sum, $3n - 150 = n$, so $2n = 150$ and $n = 75$.

121

The percentage is 77.5. Suppose a team plays a 100-game season. The players won 40 percent, or 8 games of the first 20 games (one-fifth of the season). That means the team would need to win 62 out of the next 80 games, or 77.5 percent, to win 70 percent overall.

122

Molly sold 2,500 shirts the year before.

123

To get the inverse, interchange x and y so

$$x = \frac{y-2}{y+3}.$$

Thus, $xy + 3x = y - 2$,

$$y - xy = 3x + 2,$$

$$y = \frac{3x+2}{1-x}.$$

124

Burn one candle from both ends starting at the exact same time. At that same time, start the second candle from one end only. The flames on both sides of the first candle will meet after 30 minutes. At that time, the second candle would have also burned for 30 minutes. Now light its other end, and when the flames meet, 45 minutes will have past.

125

125 cubes. $5 \times 5 \times 5 = 125$ since you would subtract 2 cubes on each face.

126

Figure (e).

The other figures have interior lines that begin and end from a vertex of the main figure. E does not.

127

Cube (e) will not look like the upper right-hand cube when unfolded.

128

Robbie did it.

Since each person could have only one false statement, and each maintained his innocence, that would be the killer's one falsehood. So anyone who blamed someone else could not be guilty — that would be a second lie. Robbie is the only one who did not blame someone else, so he is the guilty one. He used someone else's gun and Benny and Doogie are not old pals.

129

It will take them 50 minutes to meet. The two drivers covered the distance in the same time as one would have covered the 100 miles driving at 120 mph (65 mph + 55 mph).

If t is the time in hours to cover 100 miles,

$120t = 100$,

$6t = 5$, and

$t = 5/6$.

5/6 of an hour is 50 minutes.

130

In Base_4,

$$\begin{aligned} .132 &= \frac{1}{4} + \frac{3}{4^2} + \frac{2}{4^3} \\ &= \frac{16+12+2}{64} \\ &= \frac{30}{64}, \end{aligned}$$

or $\frac{15}{32}$ in Base_{10}.

131

The rule is that the number in the nth position is $n(n - 1)$, so

$0 = 1 \cdot 0$
$2 = 2 \cdot 1$
$6 = 3 \cdot 2$
$12 = 4 \cdot 3$
and so on.
The 100th term is $100 \cdot 99 = 9{,}900$.

132

With the numbers placed as shown on the diagram below, the opposite sides each will add up to seven when the cube is reformed.

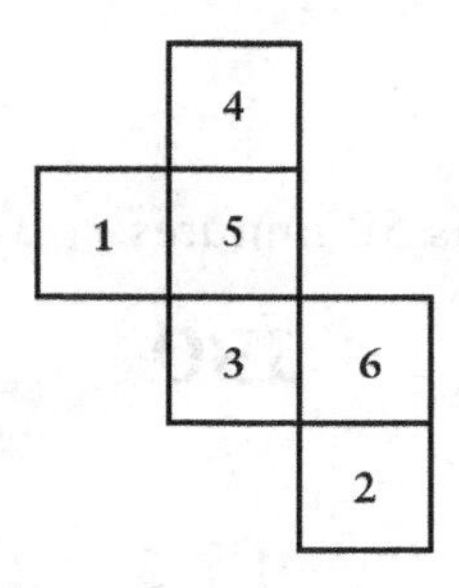

133

Move line 3:

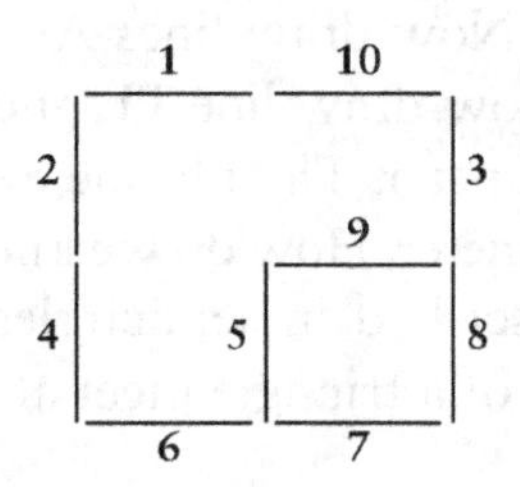

134

There are 10 birds and 20 bulls.

If a is the number of bulls and b is the number of birds, then we know that

$a + b = 30$

$4a + 2b = 100.$

Solving,

$-2a - 2b = -60$

$4a + 2b = 100$

so

$2a = 40, a = 20,$

and $b = 10.$

135

Draw line AP, making sure it intersects the circle. Do the same for BP. Now draw lines AD and BC. Call the intersection E. Now draw line PE and continue it until it touches the diameter. The PE line extended is perpendicular to the diameter. How do we know this to be true?

The angles inscribed in semicircles are right angles and the altitudes of a triangle meet at a common point.

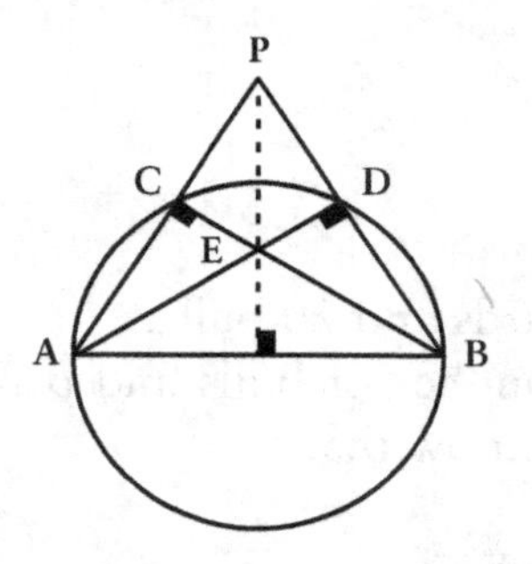

136

There are 40 divisors. A divisor of $p^4 q^7$ is $p^r q^s$, where $r = 0, 1, 2, 3,$ or 4 and $s = 0, 1, 2, 3, 4, 5, 6,$ or 7.

There are 5 choices for r and 8 choices for s, and so $5 \cdot 8 = 40$ choices for both.

137

The correct answer is (a) The right-hand weight will travel downward. The left-hand weight has two supporting sections of rope and the right-hand weight only has one. That means the left-hand weight will be exerting 1/2 as much force as the right-hand weight.

138

Forty-four steps would be visible. We know that the number of steps the escalator carries Molly is half when she takes two steps at a time. She is walking twice as fast and spending half the time.

$20 + x = y$
$16 \times 2 + x/2 = y$
$20 + x = 32 + x/2$
$40 + 2x = 64 + x$
$x = 24$
$y = 44$ steps

139

Light both ends of one of the strings (it doesn't make any difference which one). It will burn down in 20 seconds regardless of how evenly it burns. At the same instant, light one end of the second string. It will start to burn while the other is totally used up. In 20 seconds, when the first string is finished burning, light the second end of the second string. This will burn what's left in 10 more seconds, for a total of 30 seconds.

140

B would be placed sixth and H would be placed fourth. The order is

D = 1	H = 4	F = 7
E = 2	A = 5	G = 8
C = 3	B = 6	I = 9

141

44 times. The hour hand rotates twice, which means that the minute hand turns 24 times. The minute hand crosses over the hour hand 22 times. A right angle is formed twice between each of these positions, resulting in 44 times that the hands form a right angle in 24 hours.

142

The answer is $8,000 each.

Twelve people paid x dollars each for the company. Sixteen people would have paid $16(x - 2{,}000)$ each. Therefore

$$12x = 16(x - 2{,}000),$$
$$12x = 16x - 32{,}000,$$
$$4x = 32{,}000, \text{ and}$$
$$x = 8{,}000.$$

143

The answer is 22.

Here's one way to look at this:

You know z must be 1 since it's a carry-over from $5 + x$. Since z is 1, then w and y must be 6 and 7 (or 7 and 6). That means x must be 8, and

$1 + 6 + 7 + 8 = 22$.

144

The answer is (b) 10 typists.

Each typist can finish one page in five minutes.

If one typist can finish one page in five minutes, then only two pages can be completed in 10 minutes. Since we need 20 pages, we will need 10 typists to finish the 20 pages in 10 minutes.

145

The answer is 33 miles.

Here is one way to look at this:

Let x be the distance that Bibi runs. Then Bob scooters $2.5x$. The total distance $x + 2.5x = 42$, so $x = 12$ and $2.5x = 30$.

So Bibi runs at 12 miles an hour and Bob drives at 30 miles an hour. If Bob drives the scooter for 1/2 hour, he travels 15 miles. If Bibi runs for 1 1/2 hours, she travels 18 miles and $15 + 18 = 33$.

146

The missing piece is (c).

147

2/3

There are four possible outcomes, all equal in probability:

Solid added — original solid drawn.
Solid added — new solid drawn.
Stripe added — original solid drawn.
Stripe added — new stripe drawn.

Since we know a solid color was drawn from the bag, we know the outcome. In two out of the three remaining outcomes, the other ball is a solid. The probability that the ball remaining in the bag also is a solid color is 2/3.

148

$k = 3$. The sum of the second and third equations is $p + q + k(p + q) = 24$, so dividing $p + q$,

$$1 + k = \frac{24}{p+q} = \frac{24}{6} = 4$$

and $k = 3$. (And then $p = 3/2$ and $q = 9/2$.)

149

The maximum product is

$$\begin{array}{r} 631 \\ \times\, 542 \\ \hline 342{,}002. \end{array}$$

150

You would be able to trace all twelve edges.

151

The answer is 65/81 or 80.2%.

One way to look at this is to see that the man will miss the target 2/3 of the time. So $(2/3) \times (2/3) \times (2/3) \times (2/3)$ will give you the probability of no hits: 16/81.

So the probability of at least one hit is $1 - \frac{16}{81} = \frac{65}{81}$.

152

–8, 5, –3, 2, –1, 1, 0, 1, 1, 2, 3, 5, 8, 13, 21 ....

153

Figure (b) is the missing piece.

154

The answer is 20.

$$\begin{aligned} ad + be + cd + ae + bd + ce &= (a + b + c)d + (a + b + c)e \\ &= (a + b + c)(d + e) \\ &= 4 \cdot 5 = 20. \end{aligned}$$

155

The answer is 60.

Here's one way to consider this. Light 1 is hooked to two other lights. There are 10 ways to do this: 2 and 3, 3 and 4, 4 and 5, 5 and 6, 2 and 4, 3 and 5, 4 and 6, 2 and 5, 3 and 6, and 2 and 6. If you do these for all six lights, there are 60 ways.

156

The answer is 50 pounds.

At the start there were 99 pounds water, 1 pound dry potato. After a week there were x pounds water, 1 pound dry potato.

If there is 98% water,

$$\frac{x}{x+1} = .98,$$

so $x = .98x + .98$,

$.02x = .98$,

$x = 49$.

The total weight is 49 + 1 pounds.

157

$* = 5$, $\blacklozenge = 11$ and $\bigcirc = 9$

The second column has three diamonds and an asterisk, which total 38. The fourth row has three asterisks and one diamond, which total 26. So we know the following equations:

$$3\blacklozenge + 1* = 38$$
$$1\blacklozenge + 3* = 26.$$

Multiply the second equation by –3

$$3\blacklozenge + 1* = 38$$
$$-3\blacklozenge - 9* = -78$$

Adding, $-8* = -40$, so

$$* = 5,\ \blacklozenge = 11, \text{ and } \bigcirc = 9.$$

158

Yes. At least one lily pad will not have a frog on it.

159

(a) 6185

The digits of the numbers in a row add up to 20.

160

$1000 = 6(144) + 11(12) + 4(1)$, so the answer is $6y4$.

161

119 to 51 is the same as 7 to 3. For every 7 degrees change on the first thermometer, the second moves 3 degrees. If we move from 14° to 0° in the first system, the second system will be reduced by 6°. When $x = 0°$, $y = 30°$, giving us $y = 3/7x + 30$. To find where the temperatures read the same,

if $x = y = \frac{3}{7}x + 30$

then $\frac{4}{7}x = 30$,

$4x = 210$, and

$x = 52.5$.

162

The answer is 1,120.

A little inspection will reveal that 3 times the position of the number minus 2 will give you the value for that position. Value $= 3n - 2 \rightarrow (3 \times 374) - 2 = 1122 - 2$ or 1120.

163

(d) $2^{2^{-2}} = 2^{1/4} = 1.189207$.

164

Exactly 400 tickets will guarantee that one of the numbers is exactly twice as large as a previously drawn number. Try this with a smaller range of numbers, and you'll see it always comes out to two-thirds of the total!

165

Figure (b) is the missing piece from the broken cube.

166

The smaller ring will make exactly four revolutions around the larger ring. One way to see this is to realize that if both rings were rotating around each other (both spinning against each other at the same time with no slippage), the smaller ring would make three revolutions around the larger ring while the larger ring made one complete rotation. But, in our example, only one of the rings is moving, and it has an additional rotation to make, compared to the example when both are moving.

167

The number of touches is

A–3 F–7
B–3 G–4
C–5 H–3
D–3 I–4
E–6 J–3

168

Because $1/9 = 1/12 + 4/144$, the answer is $.14\ \text{Base}_{12}$.

169

(b) Always true.

170

$x = -26$.

We have $\frac{1}{3^3} \cdot 3^{59} \cdot \frac{1}{3^5} \cdot 3^{3x} = 3^{x-1}$ or

$$\frac{3^{59}}{3^3} \cdot \frac{1}{3^5} \cdot 3^{3x} = 3^{51+3x} = 3^{x-1}.$$

Since the bases are now the same, the exponents must be equal:

$51 + 3x = x - 1$

$52 = -2x$

$x = -26$.

171

The answer is (c) *a* is equal to *b*.

If $a = \frac{x}{100} y$

and $b = \frac{y}{100} x$,

then $a = b$.

172

The answer is four dots.

173

The ladder is 25 feet long. From the diagram,

$x^2 = (4/5)x^2 + 225$, so

$25x^2 = 16x^2 + 5625$ and

$x = 25$.

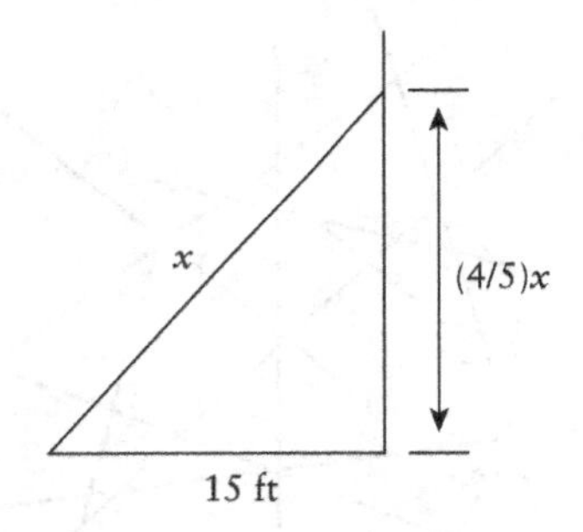

174

The Q.E.D. system is based on something other than 10. Let b represent the base of this system.

Then $4 \times 2 = 8 = b + 2$

and $5 \times 3 = 15 = 2b + 3$,

so because $2 \times 3 = 6$, then $b = 6$ in Q.E.D.

$2 \times 3 \times 4 = 24 = 4 \cdot 6 + 0$,

or 40 in Base_6.

175

S is on the face opposite H.

S is used twice.

176

There are twelve different lines. One way to see this is to number the endpoints of each line and then divide by 2.

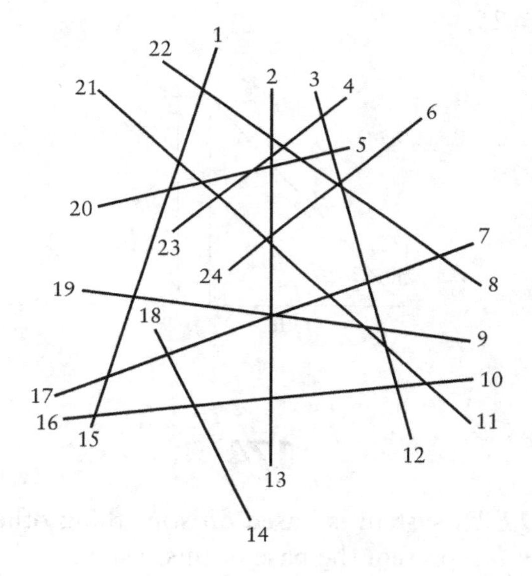

177

The answer is 8.9 mph.
Let the speed downstream be $(20 + x)$ mph.
Then the speed upstream is $(20 - x)$ mph.
We're given that

$$\frac{40}{(20+x)} + \frac{40}{(20-x)} = 5,$$

so $\frac{1600}{400 - x^2} = 5.$

Thus

$$1600 = 2000 - 5x^2,$$
$$5x^2 = 400,$$

and $x^2 = 80.$

Thus $x = 8.9$

178

Since there are 399 losers, there must be 399 matches.

179

There are sixteen missing cubes.

180

The number 72 should come next. Numbers increase by three (9, 12, 15, 18, 21, 24), and have their digits reversed.

181

The answer is (b) 39,916,800.

Since 11! includes 2, 5, and 10 that are to be multiplied together, the number must end with two zeroes.

182

The answer is six bees and five flowers.

From statement 1 you know there are more bees than flowers and from statement 2 you know there have to be at least four bees — two could be on one flower and two flowers are left out. But you can quickly see there have to be more than four bees and three flowers. You also know there is an even number of bees. Six bees and five flowers is the next choice, and it works.

183

The minimum number of square sheets of paper is 8. Imagine that the largest square is 4 × 4, its color is black, and each smaller size is a lighter shade. So, we have:

1 – 4 × 4 (black)
1 – 3 × 3 (dark gray)
3 – 2 × 2 (light gray)
3 – 1 × 1 (white)

184

The answer is (b) 1/2.

Either A throws more heads than B, or A throws more tails than B, but (since A has only one extra coin) not both. By symmetry, these two mutually exclusive possibilities occur with equal probability. The probability that A obtains more heads than B is 1/2. This probability is independent of the number of coins held by each of the players.

185

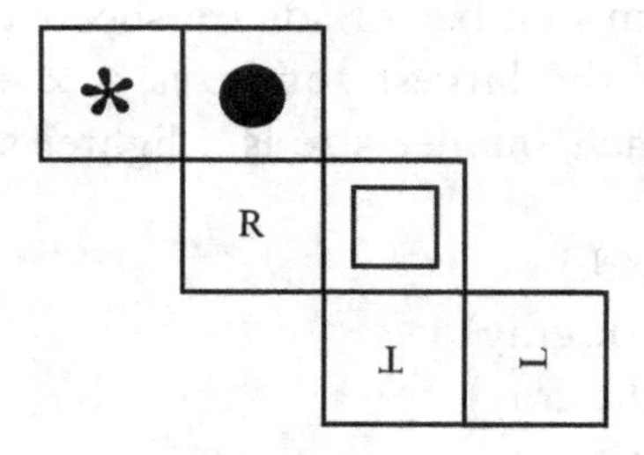

186

The answer is (c) ∞

187

It would look like (c).

188

View (d) is incorrect.

189

Seven edges of the box will need to be cut.

190

There are fourteen different calendars: seven calendars, each starting on a different day of the week, and seven more with leap years.

191

The answer is 6 units.

If x = the length of a cube, then volume is $v = x^3$. The surface area is $6x^2$.

So $x^3 = 6x^2$,

$x^3/x^2 = 6x^2/x^2$, and

$x = 6$.

192

The answer is 1512.

There are three choices for the first digit, nine choices for the second digit (can't repeat the first digit), eight choices for the third digit (can't repeat the first or second digit), and seven choices for the fourth digit (can't repeat the first, second, or third digit). So the answer is 3*9*8*7 = 1512.

193

Figure (f) does not belong.

The other figures have the square placed on the tip of a backward L.

Figure (f) forms the capital letter L.

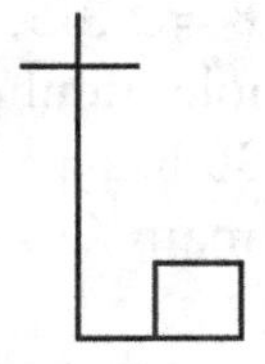

194

1. (c) is the only figure without a right angle.
2. (e) is the only figure that cannot be created without lifting the pencil or retracing any of the lines.

195

The answer is .4 pounds. There are several ways to think about this. Here is one way.

Let x be the weight of the cookies and y be the weight of the tin. Then we know

$x + y = 2$

$\frac{1}{4}x + y = .8$

or $x + y = 2$

$x + 4y = 3.2$

subtracting $-3y = -1.2$

so $y = .4$.

196

The answer is (d) 30.

The number 12 is 2·2·3. If you repeat this 100 times, you get a long chain of 200 2s and 100 3s. Now let's look at the answers: 18 = 2·3·3, which would all cancel, so $12^{100}/18$ is a whole number; 24, 27, and 36 are also all 2s and 3s. But 30 has a 5 in it that would never cancel, no matter how many 2s and 3s you have on top of the fraction.

197

The answer is 198 days.

After zero days, let the height be x. After the first day, the height is $(1 + 1/2)x$, because that equals $x + 1/2x$, which is the original height plus half of that. $(1 + 1/2)x = 3/2x$.

After the second day, the height is $(1 + 1/3)$(height after first day) $= (4/3)(3/2)x = (4/2)x$.

After the third day, the height is $(1 + 1/4)$(height after second day)$=(5/4)(4/3)(3/2)x = (5/2)x$.

So, after the nth day, the height will be $[(n+2)/2]x$; therefore, solve for n:

$100x = (n + 2)/2$[the x's cancel]

$200 = n + 2$

$n = 198$

198

The answer is 72.

Excluding the ace of spades, there are 24 arrangements of the remaining four cards. The ace of spades can be inserted into any of the three spaces between the four cards, so the answer is three times 24, or 72.

199

(a) The areas are the same.
(b) They are equal.
(c) They are equal.

200

The answer is (d), which cannot be reassembled from Fig. 1 by cutting it into two pieces.

201

One way to choose the correct path would be to ask the question:

"Would your brother say this way is the way to death?"

202

642.86 ml

Let x be the amount of salt in original 1500 ml solution.

Let y be the number of ml of water we are going to add. The new concentration has to be 70 percent of the old concentration of salt. That is,

$$\frac{x}{1500+y}=.7\left(\frac{x}{1500}\right),$$

so

$$1500x = .7x(1500 + y)$$
$$1500 = .7(1500 + y)$$
$$450 = .7y,$$

and $y = 642.86$.

203

The answer is 3 3/7 minutes.

Going downstream, Ed would go 1 1/3 miles in 4 minutes, at his current rate. Since he covers 1 mile in 4 minutes going upstream, he would go 1 1/3 + 1, or 2 1/3 miles in 8 minutes. This is 3 3/7 minutes for 1 mile.

204

The answer is (b) brother.

205

She walked 50 minutes. If she got to the office 20 minutes early and she saw the cab at *x* distance from her house, then those 20 minutes are the time it takes to cover *x* both ways. So, she came across the cab 10 minutes before her regular departure time and therefore walked 50 minutes.

206

Bill has 17 cans, Bob has 7 cans and the "certain number" they exchange is 1 can.

Let x be the number of Bill's cans, y the number of Bob's cans, and c the number Bob gives Bill.

Then we know

$x + c = 3\ (y - c)$

$x - c = 2\ (y + c)$.

Adding,

$2x = 5y - c$.

Substituting,

$$\frac{5y - c}{2} + c = 3(y - c)$$

$5y + c = 6y - 6c$

so

$y = 7c$.

So, c, the number of cans to be given to each other, must be a whole number no greater than 3. Any more than 3, and it would exceed the number of cans in a crate of pop (24).

If $c = 1$, then $y = 7$ and $x = 17$.

If Bob gives one can to Bill, Bill will have 18 and Bob is left with 6, so that works.

If Bill gives one can to Bob, Bill will have 16 and Bob will have 8. That fits as well.

207

Yes, there is a point where the distance marker will read the same for both *ents* and *tesses*. The distance markers of 50 represents both *tesses* and *ents*. Compare:

Entropy - 150 *ents*	Tesseract - 110 *tesses*
-10 *ents*	- 26 *tesses*
140 *ents*	84 *ents*

140/84 = 5/3. The ratio of 140 to 84 is 5 to 3 but we need to know how things look when Entropy is at zero. Since there are 5 *ents* for every 3 *tesses*, if we take 3/5 of any *ent* value, we will see how we compare with the *tesses* and how many more *ents* we need to add for that to be the same as *tesses*.

For example $3/5 \times 150 = 90$ but there are 110 *tesses* at 150, so we need to add 20 to the 90 to make 110.

$$3/5\ (ents) + 20 = tesses.$$

These two are the same when

$(3/5)x + 20 = x$,
$3x + 100 = 5x$,
$2x = 100$, so
$x = 50$.

208

Work backwards. I was left with one card. Let me take the extra card back from my son. Now, since I gave him half of what I had, we must have the same number of cards. I have 2, which means that he also has 2 cards. This means that I was left with 4 cards after I gave some to my daughter. Take back the extra card from my daughter. Now I have 5 cards. Since I gave her half of what I had, she must have 5 cards as well. I had 10 baseball cards in the beginning.

209

46.56 yards.

Runner B runs at .96 of the speed of Runner A and Runner C runs at .97 of the speed of Runner B so Runner C runs at .96 × .97 of the speed of Runner A, or .9312. Thus Runner C covers 93.12% of the distance in the same time.

In the race cited, Runner A completes 50 yards, and the question asked how many had Runner C completed by then.

The answer is 93.12% of 50 yards, or 46.56 yards.

At the same time (if she ran in the same race) Runner B would have completed 48 yards, and again, Runner C would have completed .97 of that distance, yielding the same answer: 46.56 yards.

210

1084 — one thousand eighty-four.

211

$-4^{-3^{-2}}$ is larger. Evaluate the stacks from top down.

-3^{-4} is $-\frac{1}{3^4}$ and $-2^{-1/81} = -.991479$.

-3^{-2} is $-\frac{1}{3^2}$ Then $-4^{(-1/9)}$ is $-.85724398$.

$-4^{-3^{-2}}$ is the larger value.

212

Bill's mother married the brother of Bill's wife.

213

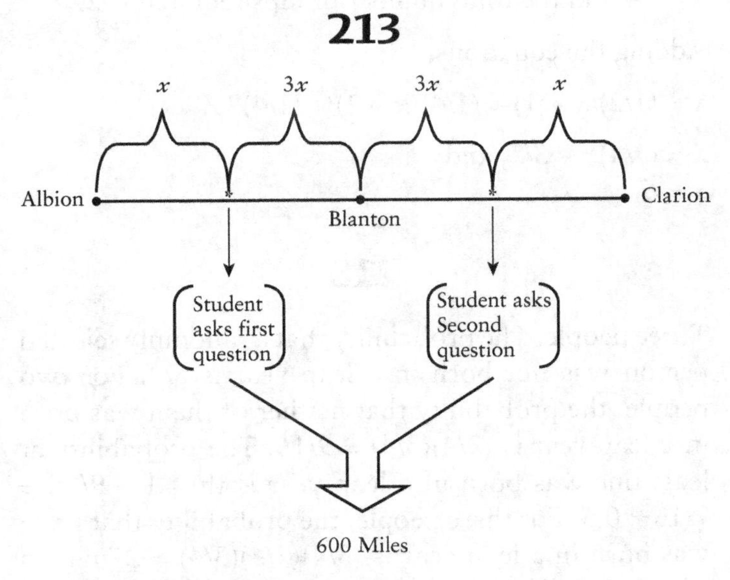

As you can see from the diagram:

$6x = 600$ miles so $x = 100$ miles.
Then $x + 3x + 3x + x = 8x$, or 800 miles.

214

6 flips. Let the expected number of coin flips be x.

1. If the first flip is tails, we have lost one flip. The probability of this event is 1/2 and the total number of flips required is $x + 1$.
2. If the first flip is heads and second flip is tails, we have lost two flips. The probability of this event is 1/4 and the total number of flips required is $x + 2$.
3. If the first flip is a heads and second flip is also heads, we are done. The probability of this event is 1/4 and the total number of flips required is 2.

Adding the equations,

$x = (1/2)(x + 1) + (1/4)(x + 2) + (1/4)2$,

$x = (3/4)x + 3/2$, and
$x = 6$

215

Three people. The probability that a randomly selected person was not born in a leap year is 3/4. For two people the probability that neither of them was born in a leap year is $(3/4)(3/4) = 9/16$. The probability at least one was born in a leap year is thus $1 - 9/16 = 7/16 < 0.5$. For three people, the probability that none was born in a leap year is $(3/4)(3/4)(3/4) = 27/64$, so the probability that at least one was born in a leap year is $1 - 27/64 = 37/64 > 0.5$. Thus a minimum of three people is needed.

216

$2^{2^{22!}}$

217

L = Horizontal	Z = 4
A = Triangle	X = 3
U = Circle	D = Vertical

△△△ = LAX

DUXDAZ = ○○○△△△△ (vertical)

A CATALOG OF SELECTED

DOVER BOOKS

IN SCIENCE AND MATHEMATICS

Astronomy

CHARIOTS FOR APOLLO: The NASA History of Manned Lunar Spacecraft to 1969, Courtney G. Brooks, James M. Grimwood, and Loyd S. Swenson, Jr. This illustrated history by a trio of experts is the definitive reference on the Apollo spacecraft and lunar modules. It traces the vehicles' design, development, and operation in space. More than 100 photographs and illustrations. 576pp. 6 3/4 x 9 1/4. 0-486-46756-2

EXPLORING THE MOON THROUGH BINOCULARS AND SMALL TELESCOPES, Ernest H. Cherrington, Jr. Informative, profusely illustrated guide to locating and identifying craters, rills, seas, mountains, other lunar features. Newly revised and updated with special section of new photos. Over 100 photos and diagrams. 240pp. 8 1/4 x 11. 0-486-24491-1

WHERE NO MAN HAS GONE BEFORE: A History of NASA's Apollo Lunar Expeditions, William David Compton. Introduction by Paul Dickson. This official NASA history traces behind-the-scenes conflicts and cooperation between scientists and engineers. The first half concerns preparations for the Moon landings, and the second half documents the flights that followed Apollo 11. 1989 edition. 432pp. 7 x 10. 0-486-47888-2

APOLLO EXPEDITIONS TO THE MOON: The NASA History, Edited by Edgar M. Cortright. Official NASA publication marks the 40th anniversary of the first lunar landing and features essays by project participants recalling engineering and administrative challenges. Accessible, jargon-free accounts, highlighted by numerous illustrations. 336pp. 8 3/8 x 10 7/8. 0-486-47175-6

ON MARS: Exploration of the Red Planet, 1958-1978--The NASA History, Edward Clinton Ezell and Linda Neuman Ezell. NASA's official history chronicles the start of our explorations of our planetary neighbor. It recounts cooperation among government, industry, and academia, and it features dozens of photos from Viking cameras. 560pp. 6 3/4 x 9 1/4. 0-486-46757-0

ARISTARCHUS OF SAMOS: The Ancient Copernicus, Sir Thomas Heath. Heath's history of astronomy ranges from Homer and Hesiod to Aristarchus and includes quotes from numerous thinkers, compilers, and scholasticists from Thales and Anaximander through Pythagoras, Plato, Aristotle, and Heraclides. 34 figures. 448pp. 5 3/8 x 8 1/2. 0-486-43886-4

AN INTRODUCTION TO CELESTIAL MECHANICS, Forest Ray Moulton. Classic text still unsurpassed in presentation of fundamental principles. Covers rectilinear motion, central forces, problems of two and three bodies, much more. Includes over 200 problems, some with answers. 437pp. 5 3/8 x 8 1/2. 0-486-64687-4

BEYOND THE ATMOSPHERE: Early Years of Space Science, Homer E. Newell. This exciting survey is the work of a top NASA administrator who chronicles technological advances, the relationship of space science to general science, and the space program's social, political, and economic contexts. 528pp. 6 3/4 x 9 1/4. 0-486-47464-X

STAR LORE: Myths, Legends, and Facts, William Tyler Olcott. Captivating retellings of the origins and histories of ancient star groups include Pegasus, Ursa Major, Pleiades, signs of the zodiac, and other constellations. "Classic." – *Sky & Telescope.* 58 illustrations. 544pp. 5 3/8 x 8 1/2. 0-486-43581-4

A COMPLETE MANUAL OF AMATEUR ASTRONOMY: Tools and Techniques for Astronomical Observations, P. Clay Sherrod with Thomas L. Koed. Concise, highly readable book discusses the selection, set-up, and maintenance of a telescope; amateur studies of the sun; lunar topography and occultations; and more. 124 figures. 26 halftones. 37 tables. 335pp. 6 1/2 x 9 1/4. 0-486-42820-6

Chemistry

MOLECULAR COLLISION THEORY, M. S. Child. This high-level monograph offers an analytical treatment of classical scattering by a central force, quantum scattering by a central force, elastic scattering phase shifts, and semi-classical elastic scattering. 1974 edition. 310pp. 5 3/8 x 8 1/2. 0-486-69437-2

HANDBOOK OF COMPUTATIONAL QUANTUM CHEMISTRY, David B. Cook. This comprehensive text provides upper-level undergraduates and graduate students with an accessible introduction to the implementation of quantum ideas in molecular modeling, exploring practical applications alongside theoretical explanations. 1998 edition. 832pp. 5 3/8 x 8 1/2. 0-486-44307-8

RADIOACTIVE SUBSTANCES, Marie Curie. The celebrated scientist's thesis, which directly preceded her 1903 Nobel Prize, discusses establishing atomic character of radioactivity; extraction from pitchblende of polonium and radium; isolation of pure radium chloride; more. 96pp. 5 3/8 x 8 1/2. 0-486-42550-9

CHEMICAL MAGIC, Leonard A. Ford. Classic guide provides intriguing entertainment while elucidating sound scientific principles, with more than 100 unusual stunts: cold fire, dust explosions, a nylon rope trick, a disappearing beaker, much more. 128pp. 5 3/8 x 8 1/2. 0-486-67628-5

ALCHEMY, E. J. Holmyard. Classic study by noted authority covers 2,000 years of alchemical history: religious, mystical overtones; apparatus; signs, symbols, and secret terms; advent of scientific method, much more. Illustrated. 320pp. 5 3/8 x 8 1/2. 0-486-26298-7

CHEMICAL KINETICS AND REACTION DYNAMICS, Paul L. Houston. This text teaches the principles underlying modern chemical kinetics in a clear, direct fashion, using several examples to enhance basic understanding. Solutions to selected problems. 2001 edition. 352pp. 8 3/8 x 11. 0-486-45334-0

PROBLEMS AND SOLUTIONS IN QUANTUM CHEMISTRY AND PHYSICS, Charles S. Johnson and Lee G. Pedersen. Unusually varied problems, with detailed solutions, cover of quantum mechanics, wave mechanics, angular momentum, molecular spectroscopy, scattering theory, more. 280 problems, plus 139 supplementary exercises. 430pp. 6 1/2 x 9 1/4. 0-486-65236-X

ELEMENTS OF CHEMISTRY, Antoine Lavoisier. Monumental classic by the founder of modern chemistry features first explicit statement of law of conservation of matter in chemical change, and more. Facsimile reprint of original (1790) Kerr translation. 539pp. 5 3/8 x 8 1/2. 0-486-64624-6

MAGNETISM AND TRANSITION METAL COMPLEXES, F. E. Mabbs and D. J. Machin. A detailed view of the calculation methods involved in the magnetic properties of transition metal complexes, this volume offers sufficient background for original work in the field. 1973 edition. 240pp. 5 3/8 x 8 1/2. 0-486-46284-6

GENERAL CHEMISTRY, Linus Pauling. Revised third edition of classic first-year text by Nobel laureate. Atomic and molecular structure, quantum mechanics, statistical mechanics, thermodynamics correlated with descriptive chemistry. Problems. 992pp. 5 3/8 x 8 1/2. 0-486-65622-5

ELECTROLYTE SOLUTIONS: Second Revised Edition, R. A. Robinson and R. H. Stokes. Classic text deals primarily with measurement, interpretation of conductance, chemical potential, and diffusion in electrolyte solutions. Detailed theoretical interpretations, plus extensive tables of thermodynamic and transport properties. 1970 edition. 590pp. 5 3/8 x 8 1/2. 0-486-42225-9

Engineering

FUNDAMENTALS OF ASTRODYNAMICS, Roger R. Bate, Donald D. Mueller, and Jerry E. White. Teaching text developed by U.S. Air Force Academy develops the basic two-body and n-body equations of motion; orbit determination; classical orbital elements, coordinate transformations; differential correction; more. 1971 edition. 455pp. 5 3/8 x 8 1/2. 0-486-60061-0

INTRODUCTION TO CONTINUUM MECHANICS FOR ENGINEERS: Revised Edition, Ray M. Bowen. This self-contained text introduces classical continuum models within a modern framework. Its numerous exercises illustrate the governing principles, linearizations, and other approximations that constitute classical continuum models. 2007 edition. 320pp. 6 1/8 x 9 1/4. 0-486-47460-7

ENGINEERING MECHANICS FOR STRUCTURES, Louis L. Bucciarelli. This text explores the mechanics of solids and statics as well as the strength of materials and elasticity theory. Its many design exercises encourage creative initiative and systems thinking. 2009 edition. 320pp. 6 1/8 x 9 1/4. 0-486-46855-0

FEEDBACK CONTROL THEORY, John C. Doyle, Bruce A. Francis and Allen R. Tannenbaum. This excellent introduction to feedback control system design offers a theoretical approach that captures the essential issues and can be applied to a wide range of practical problems. 1992 edition. 224pp. 6 1/2 x 9 1/4. 0-486-46933-6

THE FORCES OF MATTER, Michael Faraday. These lectures by a famous inventor offer an easy-to-understand introduction to the interactions of the universe's physical forces. Six essays explore gravitation, cohesion, chemical affinity, heat, magnetism, and electricity. 1993 edition. 96pp. 5 3/8 x 8 1/2. 0-486-47482-8

DYNAMICS, Lawrence E. Goodman and William H. Warner. Beginning engineering text introduces calculus of vectors, particle motion, dynamics of particle systems and plane rigid bodies, technical applications in plane motions, and more. Exercises and answers in every chapter. 619pp. 5 3/8 x 8 1/2. 0-486-42006-X

ADAPTIVE FILTERING PREDICTION AND CONTROL, Graham C. Goodwin and Kwai Sang Sin. This unified survey focuses on linear discrete-time systems and explores natural extensions to nonlinear systems. It emphasizes discrete-time systems, summarizing theoretical and practical aspects of a large class of adaptive algorithms. 1984 edition. 560pp. 6 1/2 x 9 1/4. 0-486-46932-8

INDUCTANCE CALCULATIONS, Frederick W. Grover. This authoritative reference enables the design of virtually every type of inductor. It features a single simple formula for each type of inductor, together with tables containing essential numerical factors. 1946 edition. 304pp. 5 3/8 x 8 1/2. 0-486-47440-2

THERMODYNAMICS: Foundations and Applications, Elias P. Gyftopoulos and Gian Paolo Beretta. Designed by two MIT professors, this authoritative text discusses basic concepts and applications in detail, emphasizing generality, definitions, and logical consistency. More than 300 solved problems cover realistic energy systems and processes. 800pp. 6 1/8 x 9 1/4. 0-486-43932-1

THE FINITE ELEMENT METHOD: Linear Static and Dynamic Finite Element Analysis, Thomas J. R. Hughes. Text for students without in-depth mathematical training, this text includes a comprehensive presentation and analysis of algorithms of time-dependent phenomena plus beam, plate, and shell theories. Solution guide available upon request. 672pp. 6 1/2 x 9 1/4. 0-486-41181-8

Mathematics–Bestsellers

HANDBOOK OF MATHEMATICAL FUNCTIONS: with Formulas, Graphs, and Mathematical Tables, Edited by Milton Abramowitz and Irene A. Stegun. A classic resource for working with special functions, standard trig, and exponential logarithmic definitions and extensions, it features 29 sets of tables, some to as high as 20 places. 1046pp. 8 x 10 1/2. 0-486-61272-4

ABSTRACT AND CONCRETE CATEGORIES: The Joy of Cats, Jiri Adamek, Horst Herrlich, and George E. Strecker. This up-to-date introductory treatment employs category theory to explore the theory of structures. Its unique approach stresses concrete categories and presents a systematic view of factorization structures. Numerous examples. 1990 edition, updated 2004. 528pp. 6 1/8 x 9 1/4. 0-486-46934-4

MATHEMATICS: Its Content, Methods and Meaning, A. D. Aleksandrov, A. N. Kolmogorov, and M. A. Lavrent'ev. Major survey offers comprehensive, coherent discussions of analytic geometry, algebra, differential equations, calculus of variations, functions of a complex variable, prime numbers, linear and non-Euclidean geometry, topology, functional analysis, more. 1963 edition. 1120pp. 5 3/8 x 8 1/2. 0-486-40916-3

INTRODUCTION TO VECTORS AND TENSORS: Second Edition--Two Volumes Bound as One, Ray M. Bowen and C.-C. Wang. Convenient single-volume compilation of two texts offers both introduction and in-depth survey. Geared toward engineering and science students rather than mathematicians, it focuses on physics and engineering applications. 1976 edition. 560pp. 6 1/2 x 9 1/4. 0-486-46914-X

AN INTRODUCTION TO ORTHOGONAL POLYNOMIALS, Theodore S. Chihara. Concise introduction covers general elementary theory, including the representation theorem and distribution functions, continued fractions and chain sequences, the recurrence formula, special functions, and some specific systems. 1978 edition. 272pp. 5 3/8 x 8 1/2. 0-486-47929-3

ADVANCED MATHEMATICS FOR ENGINEERS AND SCIENTISTS, Paul DuChateau. This primary text and supplemental reference focuses on linear algebra, calculus, and ordinary differential equations. Additional topics include partial differential equations and approximation methods. Includes solved problems. 1992 edition. 400pp. 7 1/2 x 9 1/4. 0-486-47930-7

PARTIAL DIFFERENTIAL EQUATIONS FOR SCIENTISTS AND ENGINEERS, Stanley J. Farlow. Practical text shows how to formulate and solve partial differential equations. Coverage of diffusion-type problems, hyperbolic-type problems, elliptic-type problems, numerical and approximate methods. Solution guide available upon request. 1982 edition. 414pp. 6 1/8 x 9 1/4. 0-486-67620-X

VARIATIONAL PRINCIPLES AND FREE-BOUNDARY PROBLEMS, Avner Friedman. Advanced graduate-level text examines variational methods in partial differential equations and illustrates their applications to free-boundary problems. Features detailed statements of standard theory of elliptic and parabolic operators. 1982 edition. 720pp. 6 1/8 x 9 1/4. 0-486-47853-X

LINEAR ANALYSIS AND REPRESENTATION THEORY, Steven A. Gaal. Unified treatment covers topics from the theory of operators and operator algebras on Hilbert spaces; integration and representation theory for topological groups; and the theory of Lie algebras, Lie groups, and transform groups. 1973 edition. 704pp. 6 1/8 x 9 1/4. 0-486-47851-3